Ready Notes

for use with

Ecology
Concepts and Applications

Manuel C. Molles, Jr.
University of New Mexico

The McGraw-Hill Companies, Inc
Primis Custom Publishing

New York St. Louis San Francisco Auckland Bogota
Caracas Lisbon London Madrid Mexico Milan Montreal
New Delhi Paris San Juan Singapore Sydney Tokyo Toronto

McGraw-Hill
A Division of The McGraw-Hill Companies

Copyright © 1999 by The McGraw-Hill Companies, Inc. All rights reserved. Printed in the United States of America. Except as permitted under the United States Copyright Act of 1976, no part of this publication may be reproduced or distributed in any form or by any means, or stored in a data base retrieval system, without prior written permission of the publisher.

McGraw-Hill's College Custom Series consists of products that are produced from camera-ready copy. Peer review, class testing, and accuracy are primarily the responsibility of the author(s).

1 2 3 4 5 6 7 8 9 0   QPD   0 9

ISBN 0-07-235478-X

Welcome to

READY NOTES

Your life just got easier! This booklet includes *Ready Notes* to accompany your textbook. *Ready Notes* were designed as a classroom supplement to accompany *Ready Shows*. More importantly, *Ready Notes* were developed for you, the student.

Somewhere in your educational experience, you have undoubtedly encountered a common dilemma facing many students; the feeling of helplessness that comes from trying to write down everything your instructor says and, at the same time, actually paying attention to what is being taught. *Ready Notes* addresses this problem by providing pre-prepared lecture outlines to accompany the *Ready Shows* your instructor will be using in class. Rather than spending time copying material that is already in the book, you will be able to focus on the most important aspects of what your instructor is actually saying. You will still be expected to take notes, but the nature of those notes will change.

Each page in *Ready Notes* includes reproductions of some of the actual projected screens that you will be seeing in class. The *Ready Notes* booklet includes the information for many of the examples that your instructor will be presenting.

It is your responsibility to attend class regularly and to be prepared for class. However, used properly, *Ready Notes* will help you to achieve your goals for the course. Good luck!

CONTENTS

1	Introduction: What Is Ecology?	1
2	Life on Land	5
3	Life in Water	15
4	Temperature Relations	25
5	Water Relations	37
6	Energy and Nutrient Relations	47
7	Population Distribution and Abundance	57
8	Population Dynamics	65
9	Population Growth	73
10	Competition	83
11	Exploitation: Predation, Herbivory, Parasitism, and Disease	91
12	Mutualism	99
13	Species Abundance and Diversity	105
14	Food Webs	113
15	Primary Production and Energy Flow	121
16	Nutrient Cycling and Retention	129
17	Succession and Stability	139
18	Landscape Ecology	149
19	Geographic Ecology	157
20	Global Ecology	167

Chapter 1

Ecology
Concepts and Applications
First Edition

Manuel C. Molles, Jr.

(c) The McGraw-Hill Companies, Inc.

Figure 1.1

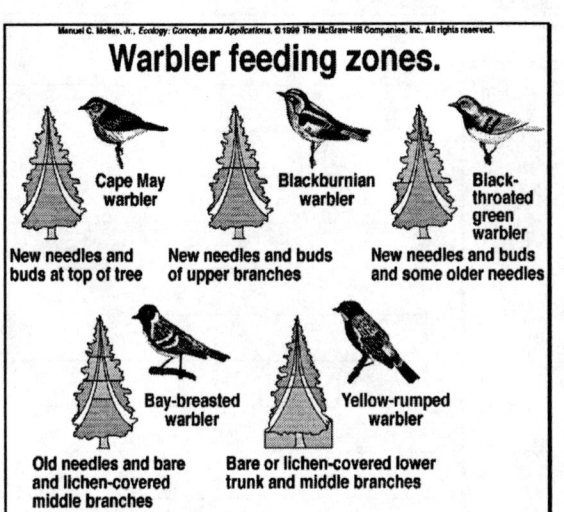

Figure 1.2

Figure 1.3

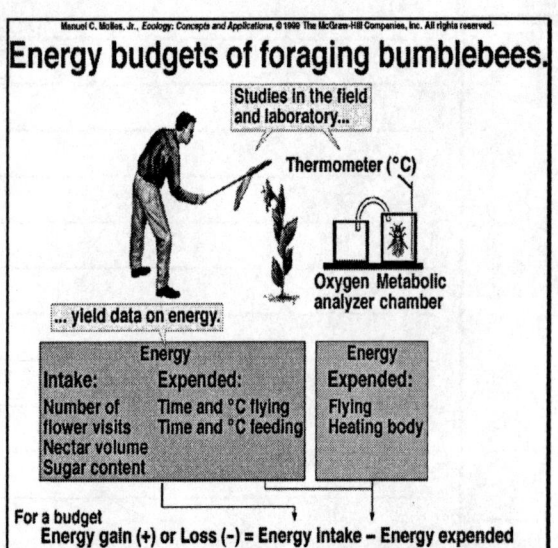

Figure 1.4

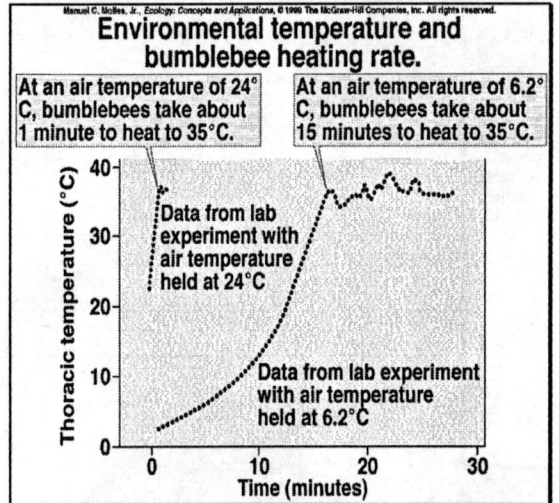

Figure 1.6

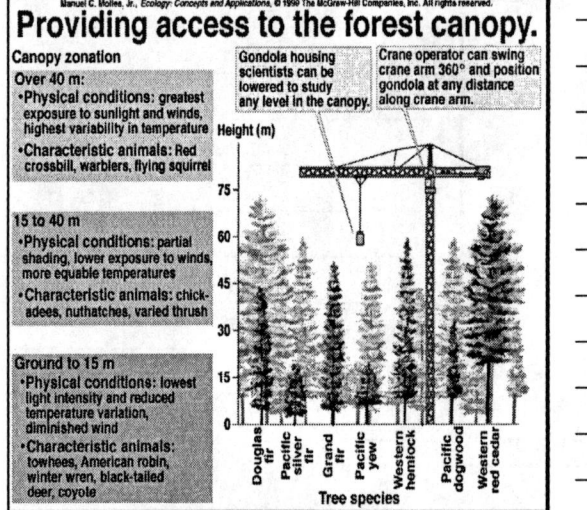

Figure 1.7

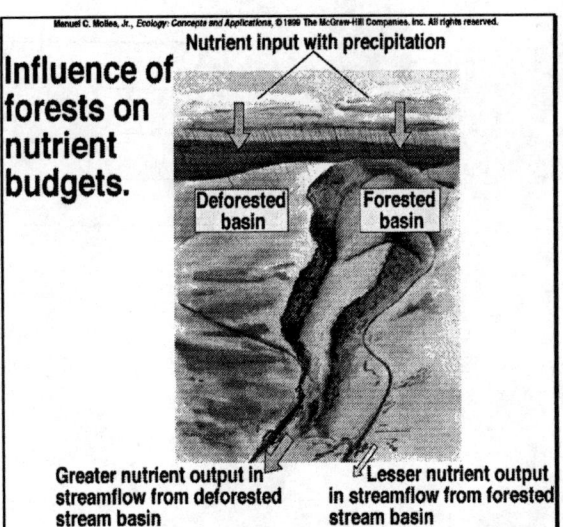

Figure 1.8

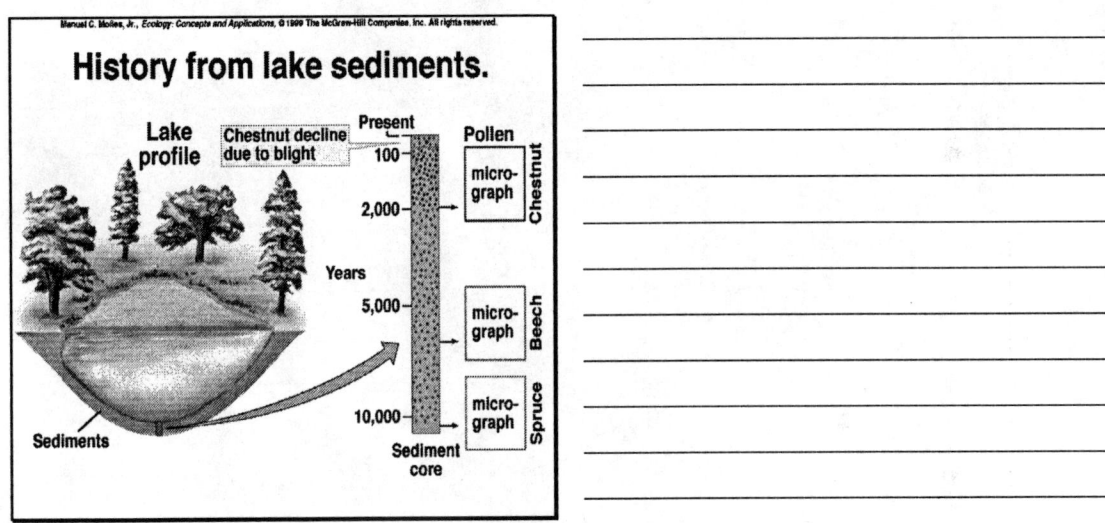

Figure 2.2

Figure 2.3

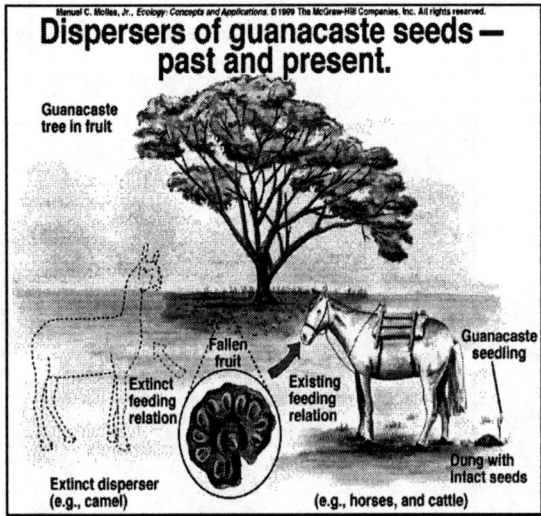

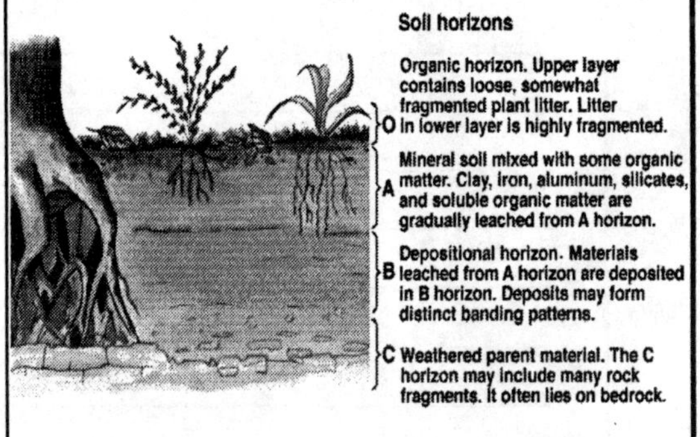

Leaching caused by rain

Red soil in Southern US have fewer nutrients

Figure 2.4

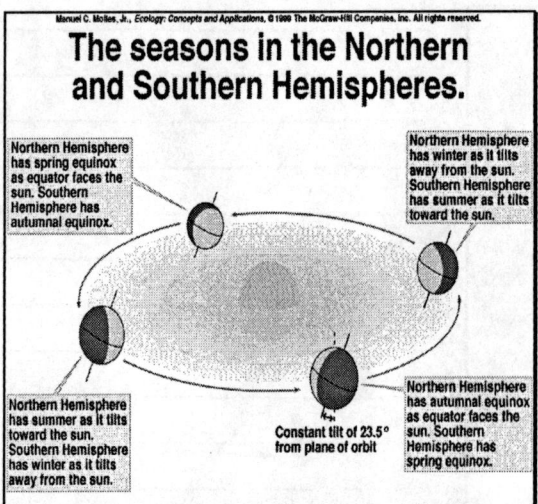

Figure 2.5

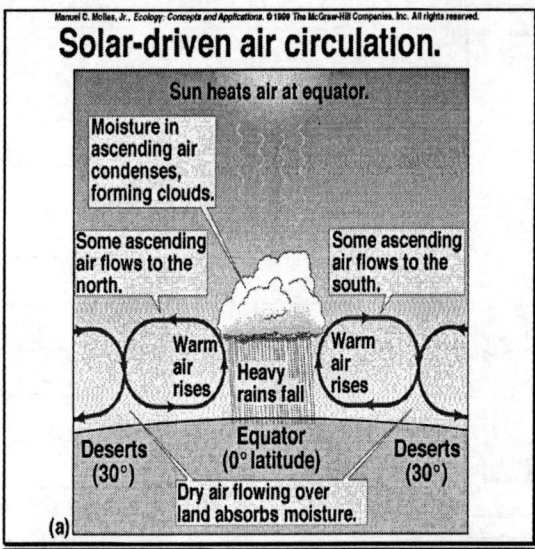

Figure 2.5

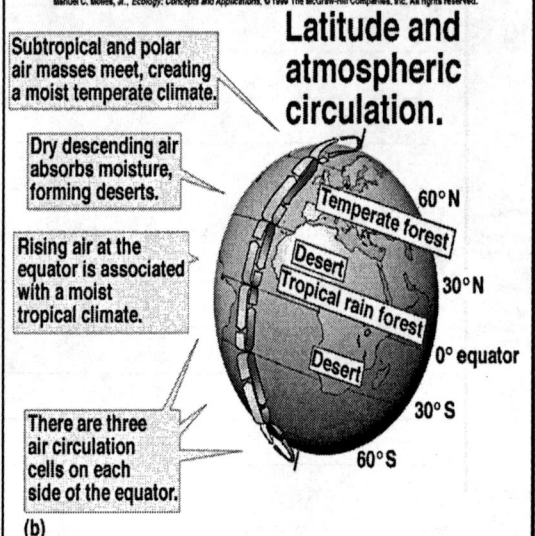

Figure 2.6

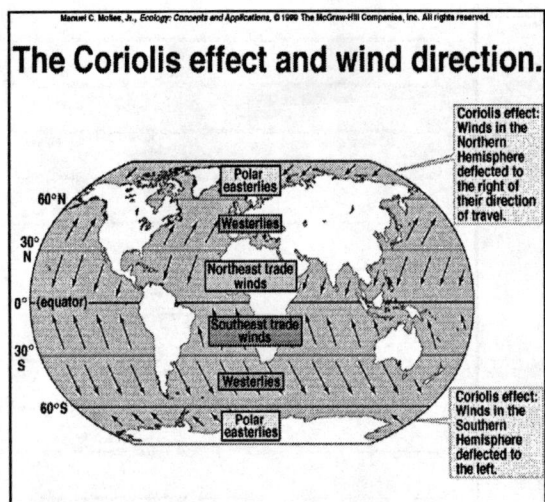

Figure 2.7

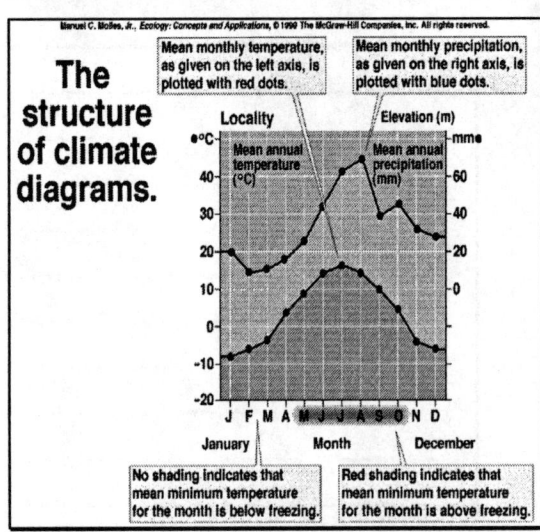

Figure 2.8

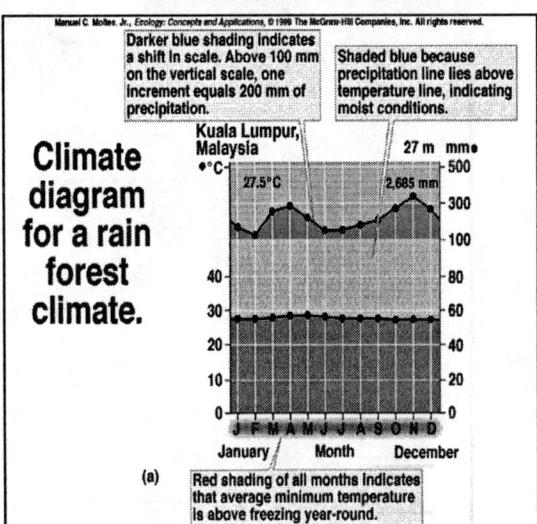

Figure 2.8

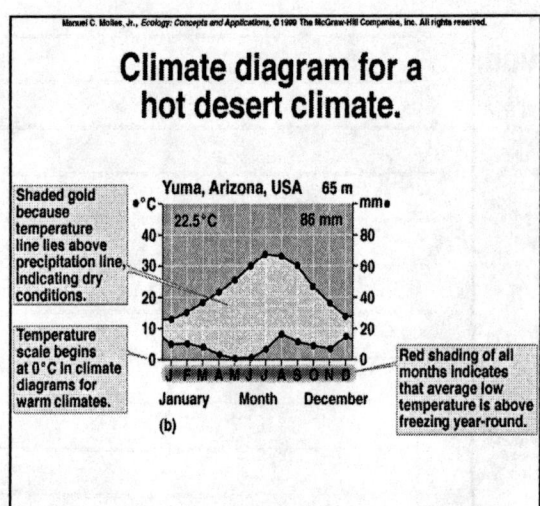

Figure 2.8

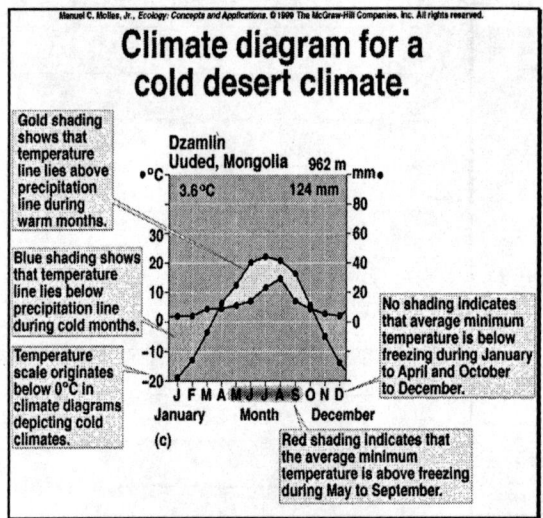

Figure 2.10

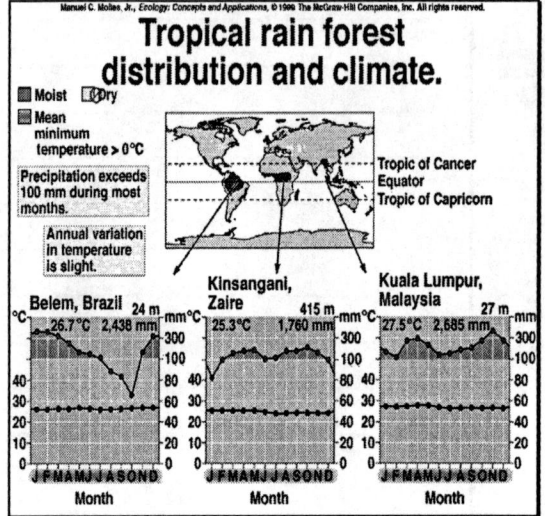

Figure 2.11

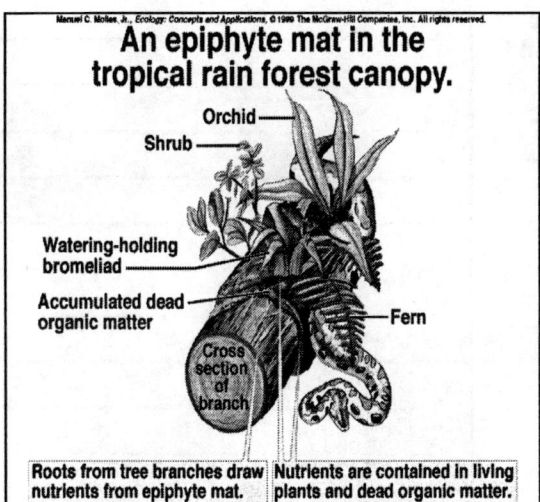

Figure 2.13

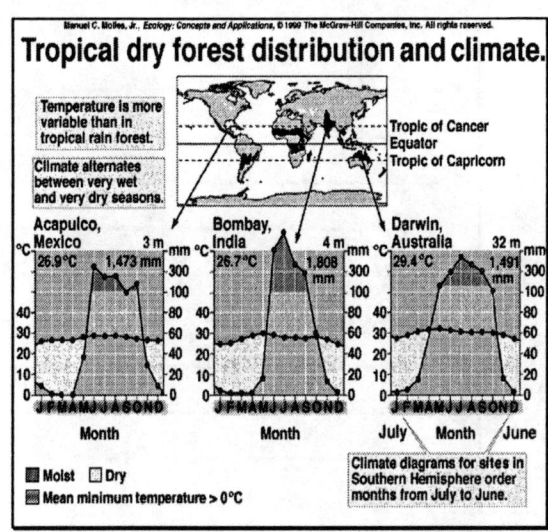

Figure 2.14

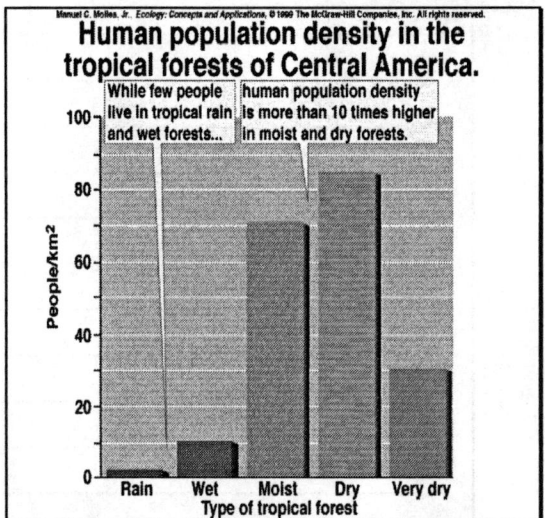

Figure 2.16

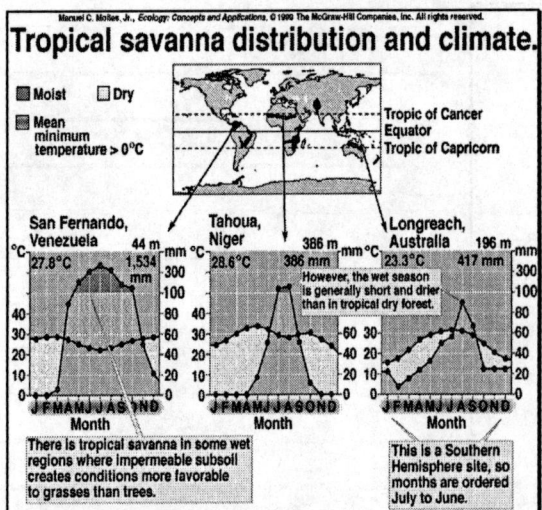

Figure 2.19

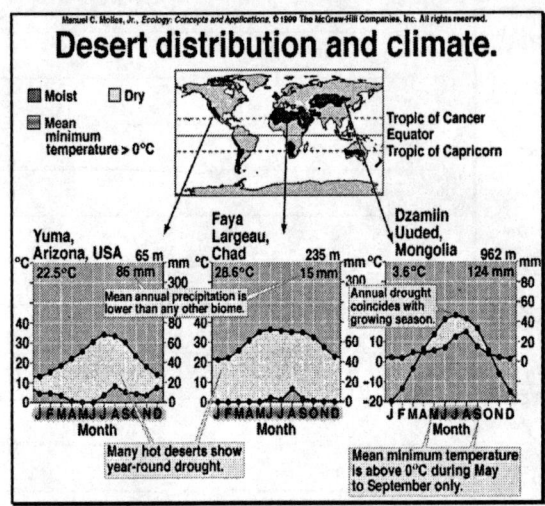

Figure 2.22

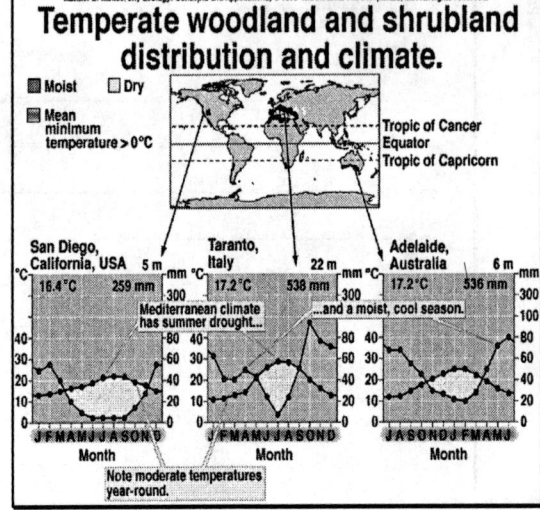

Figure 2.25

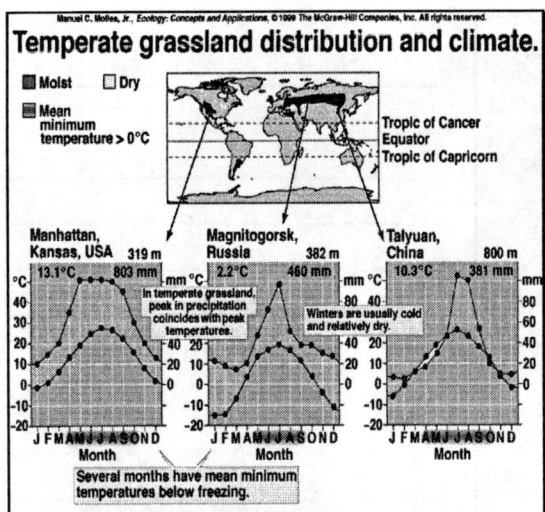

Figure 2.28

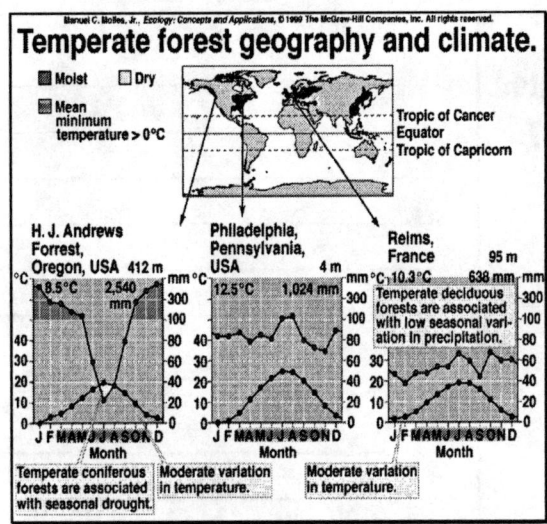

Figure 2.31

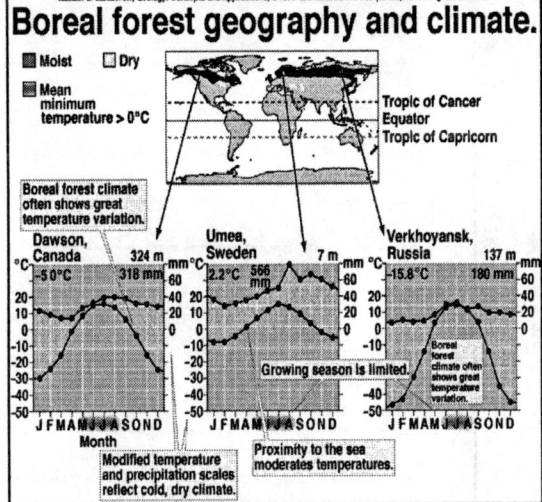

Figure 2.34
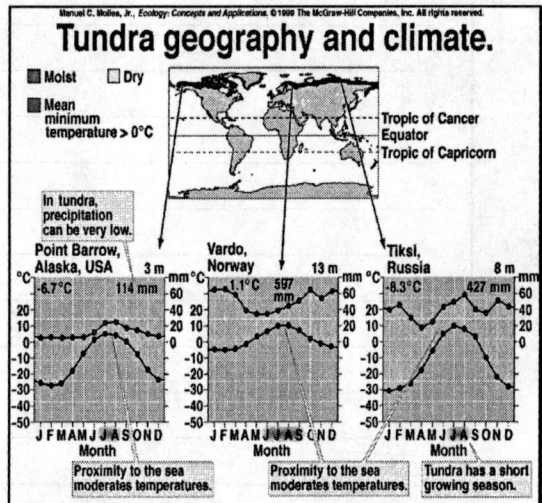

Figure 2.38
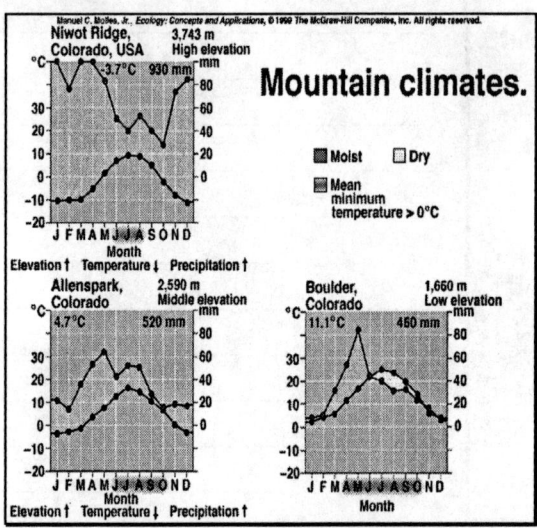

Figure 2.40
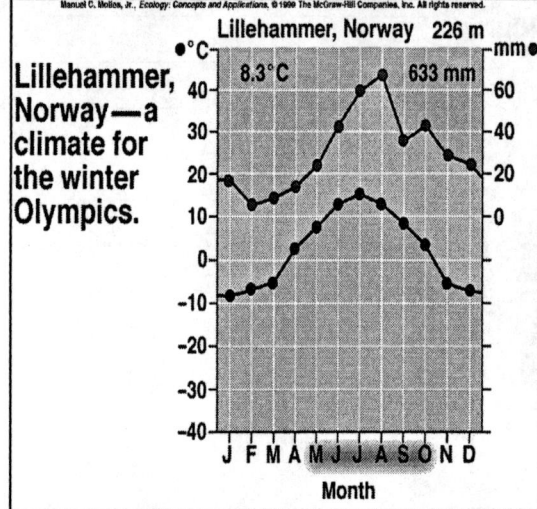

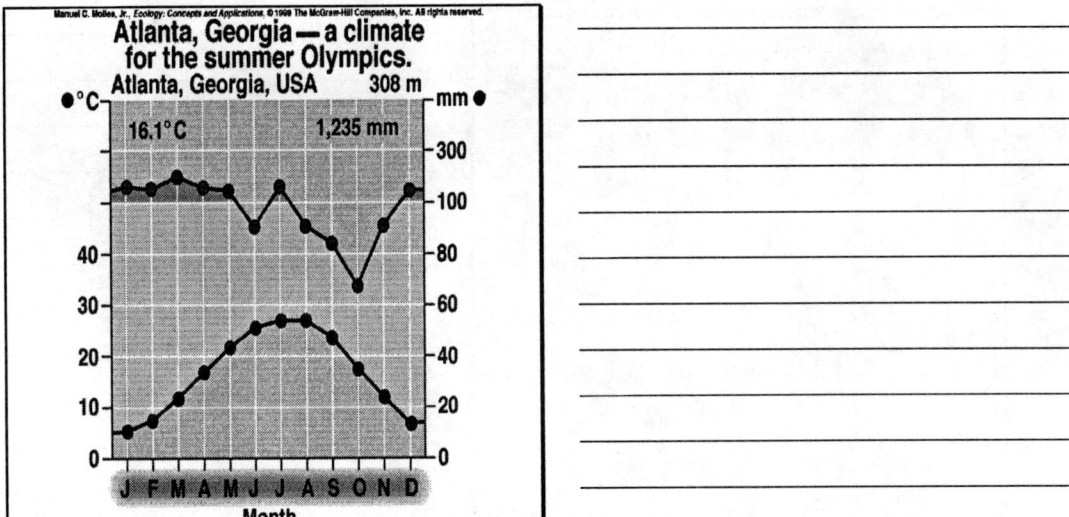

Chapter 3

Ecology
Concepts and Applications
First Edition

Manuel C. Molles, Jr.

(c) The McGraw-Hill Companies, Inc.

Figure 3.2

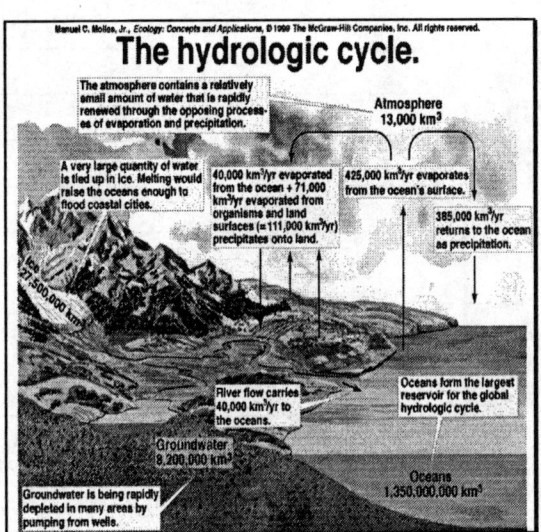

Figure 3.4

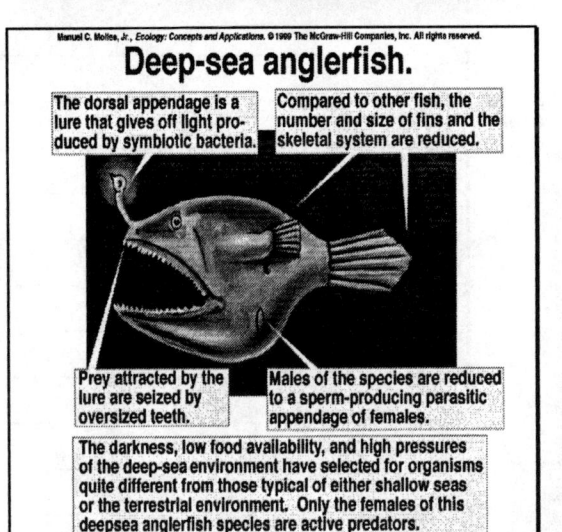

Figure 3.5

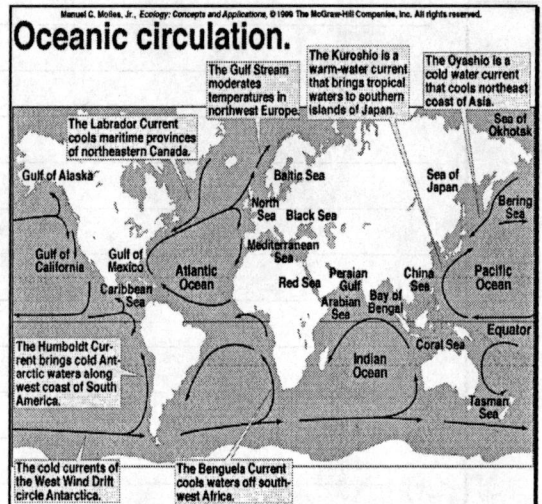

Figure 3.6

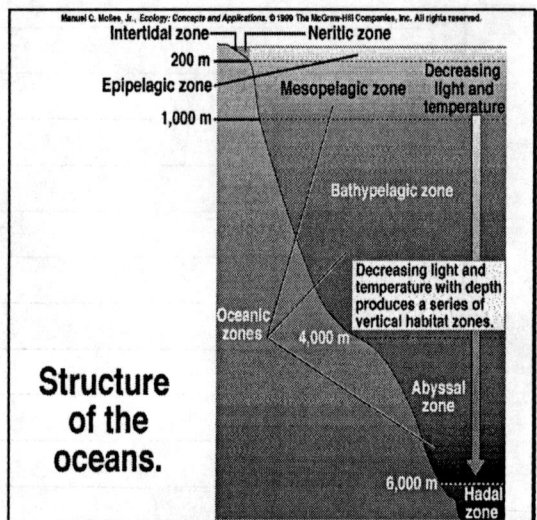

pelagic = open ocean

Figure 3.9

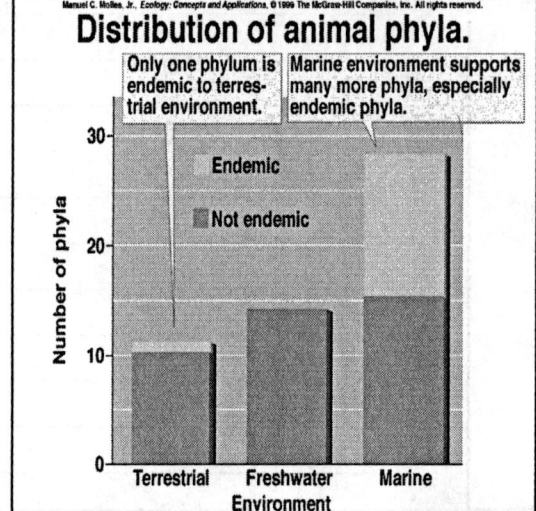

Figure 3.12

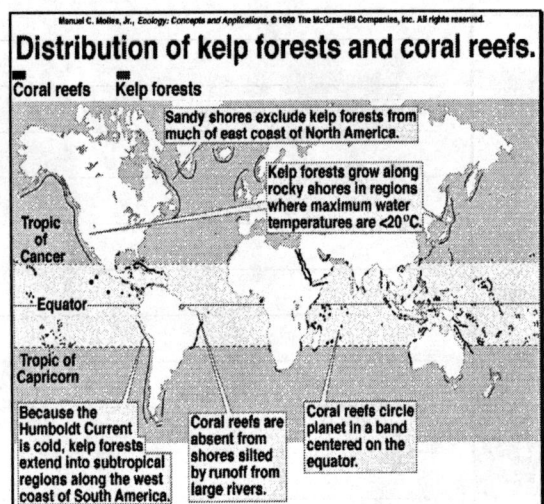

Figure 3.13

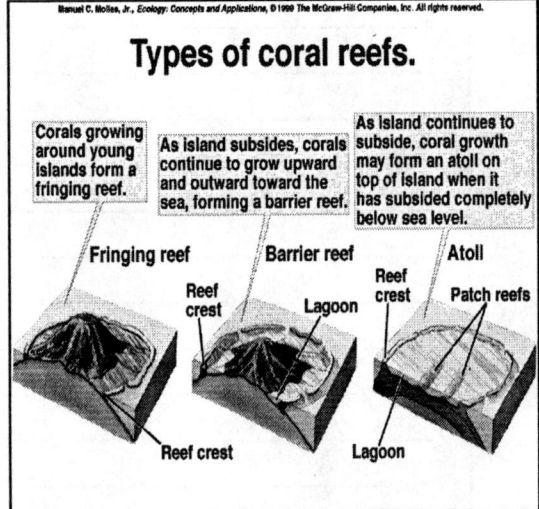

Figure 3.14

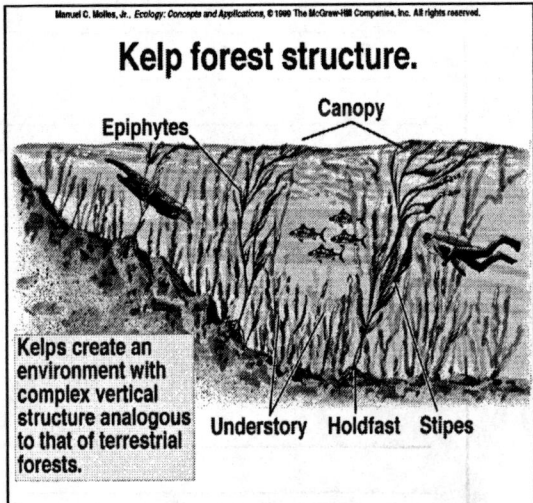

Figure 3.17

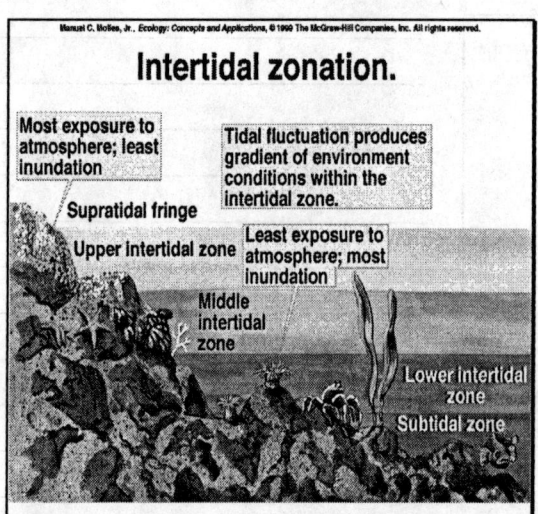

Figure 3.22

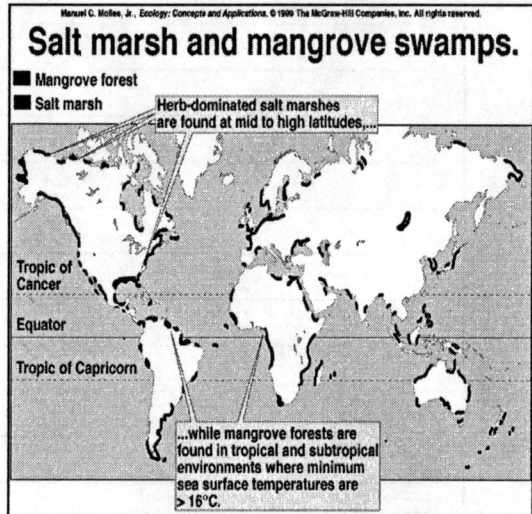

Figure 3.24

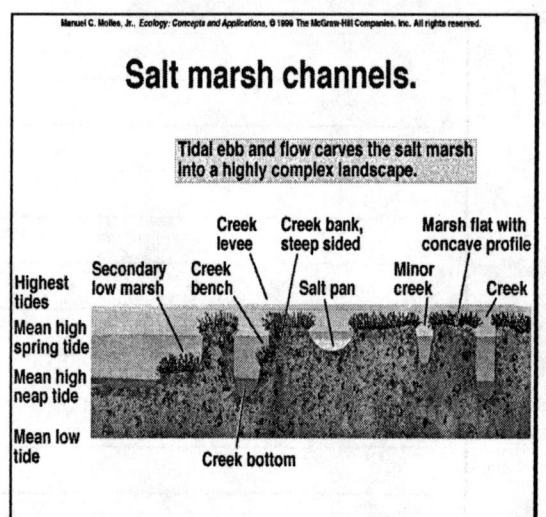

Figure 3.25

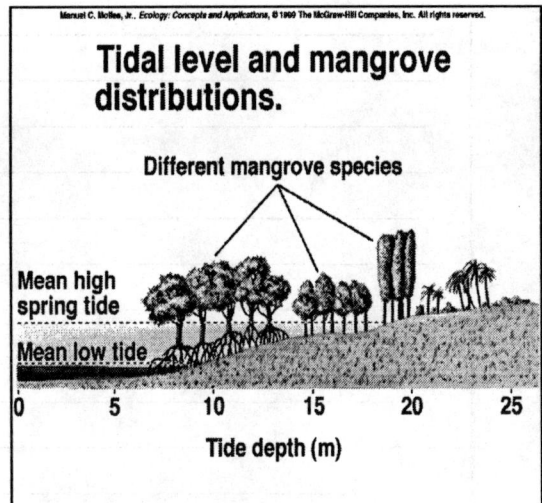

Figure 3.26

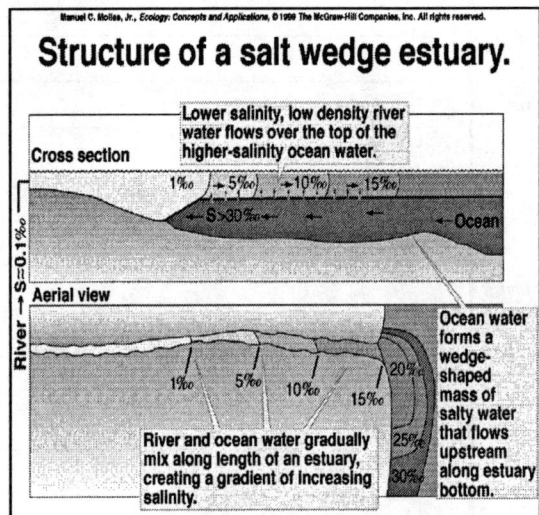

Figure 3.29

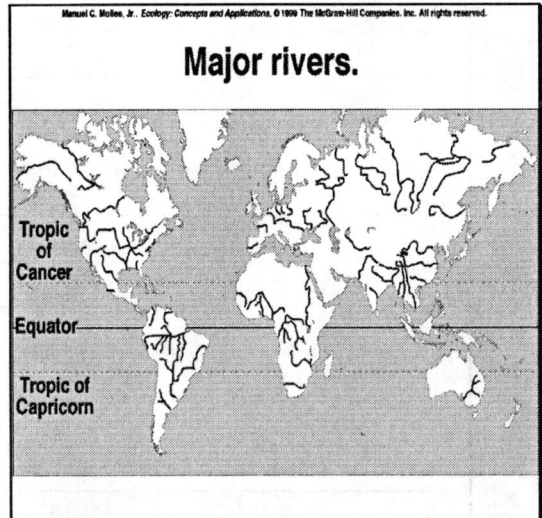

Figure 3.30

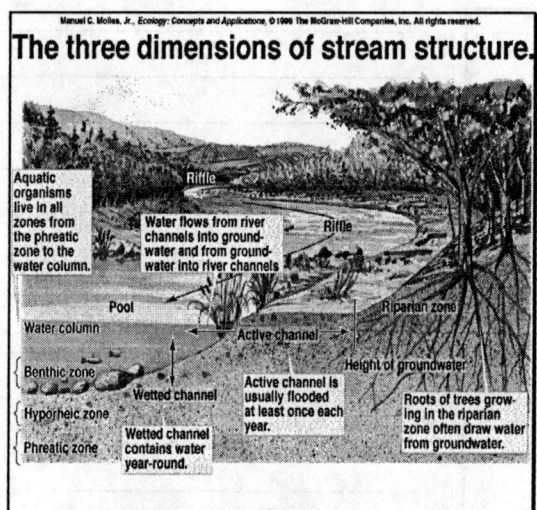

Figure 3.32

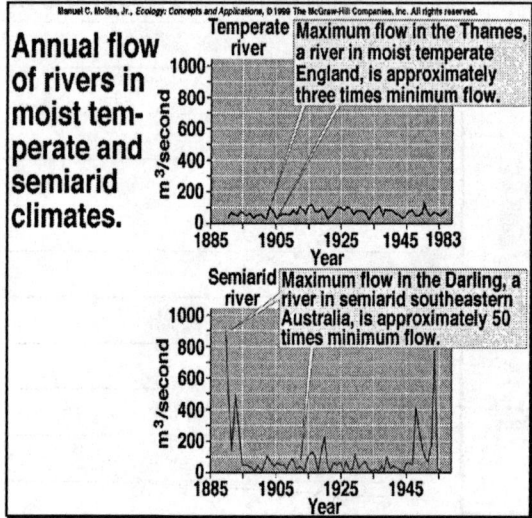

Figure 3.33

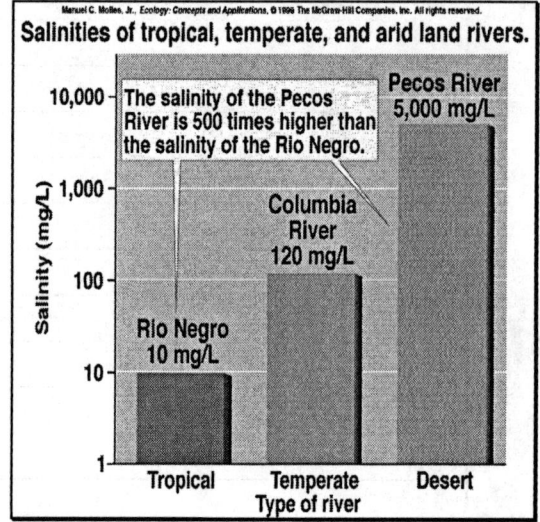

Figure 3.34

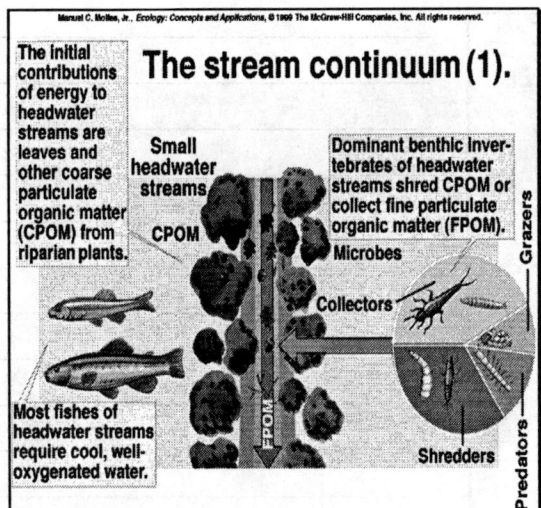

Figure 3.34

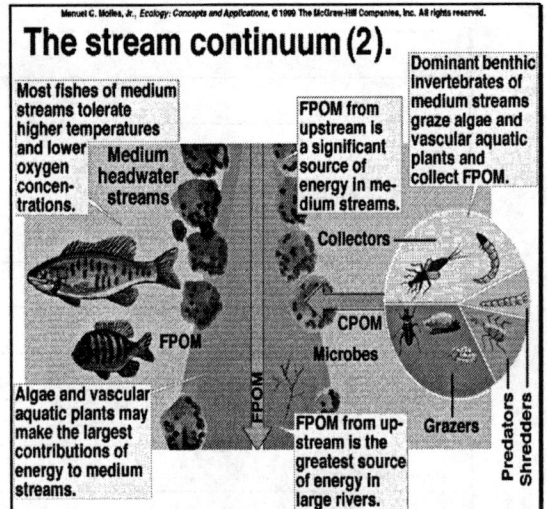

Figure 3.34

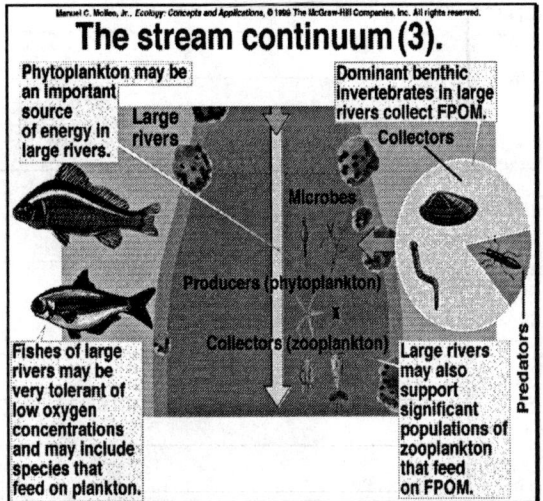

Figure 3.36
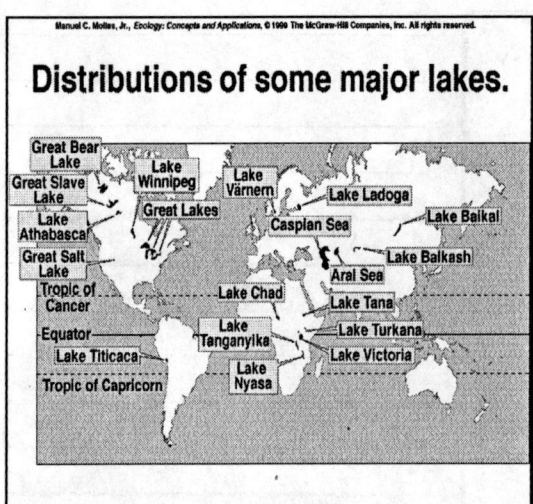

Figure 3.37
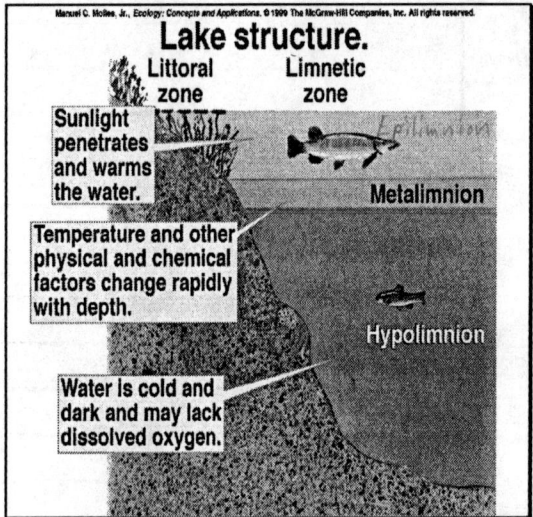

Figure 3.38
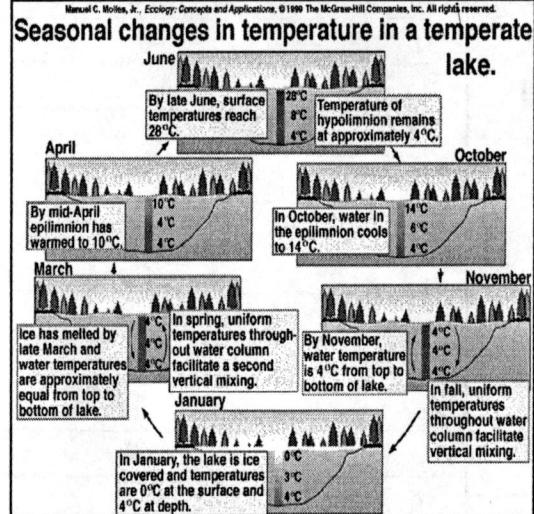

Figure 3.39

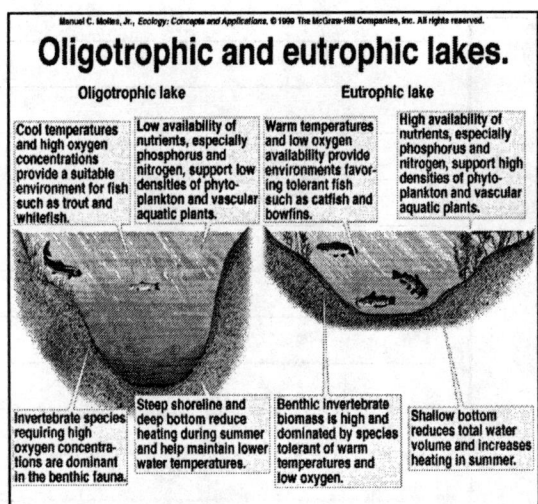

Figure 3.40

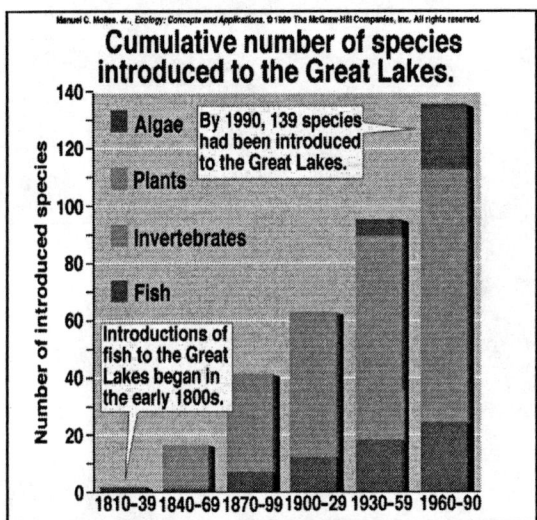

Figure 3.42

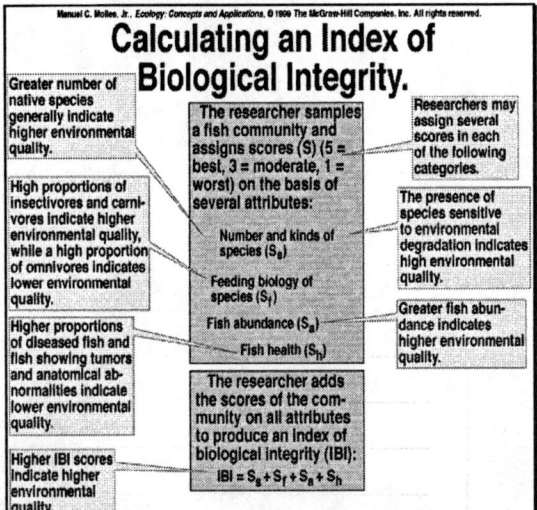

Figure 3.43

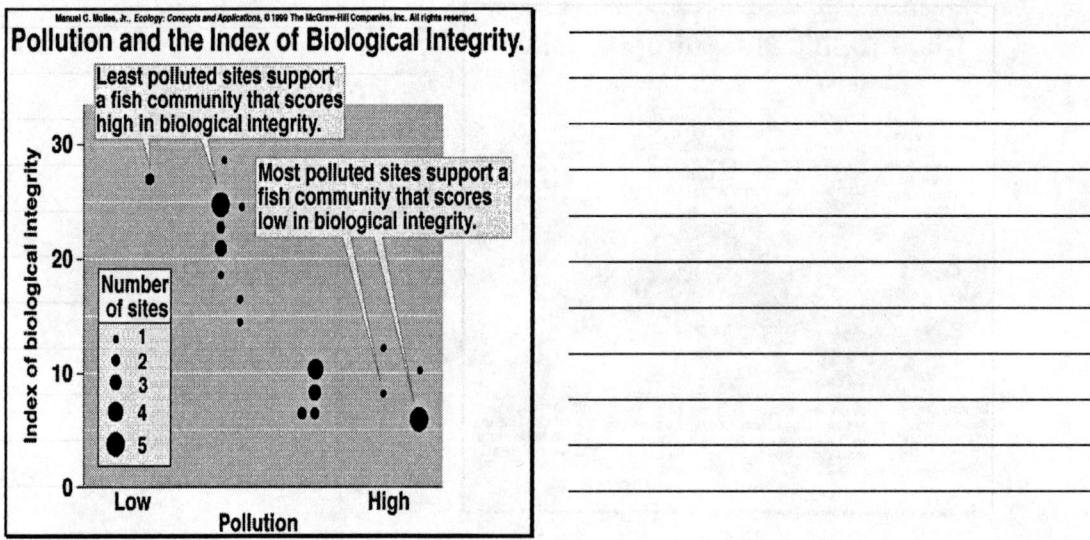

Chapter 4

Ecology

Concepts and Applications
First Edition

Manuel C. Molles, Jr.

(c) The McGraw-Hill Companies, Inc.

Figure 4.1

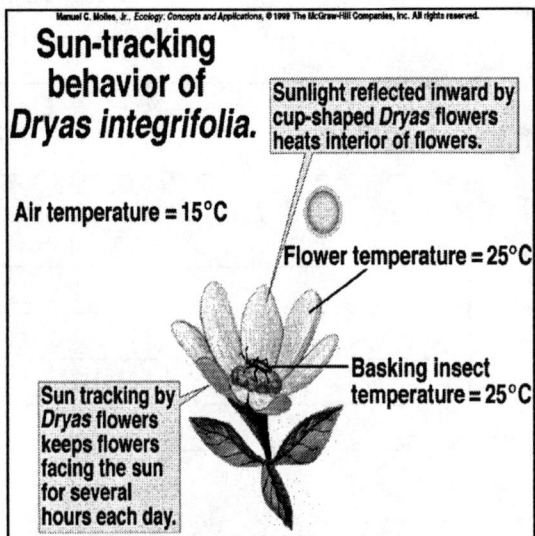

Figure 4.3

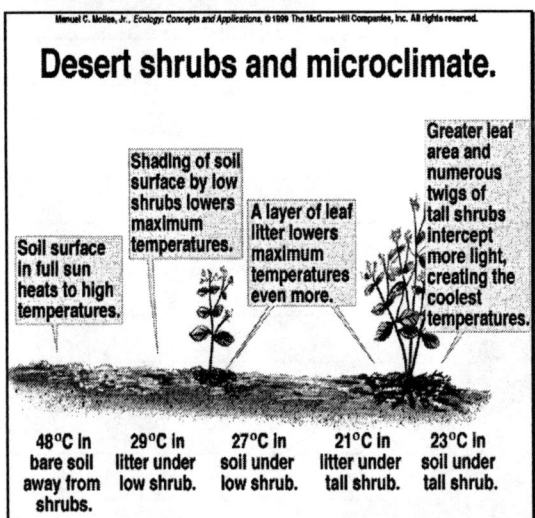

Figure 4.5

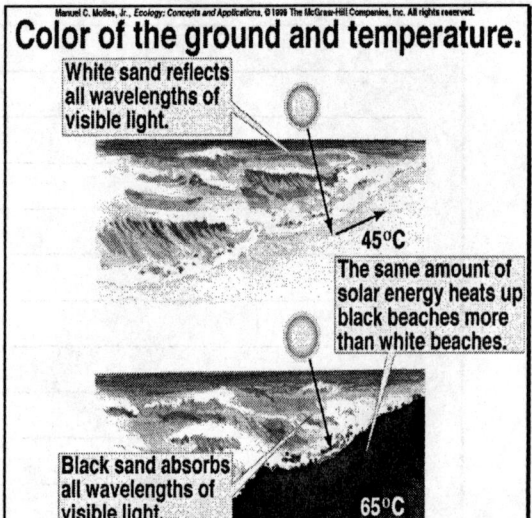

Figure 4.6

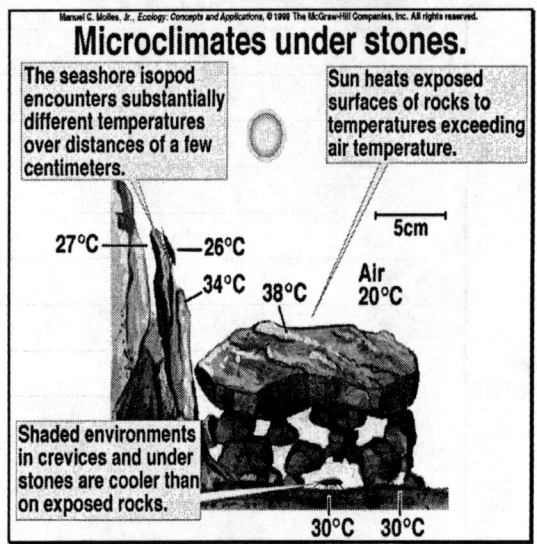

Figure 4.7

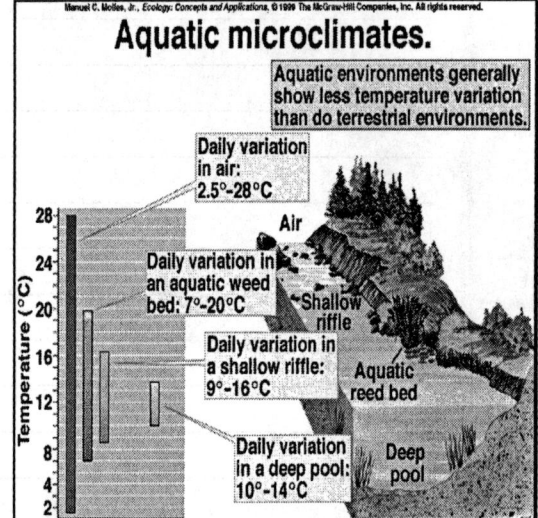

Figure 4.8

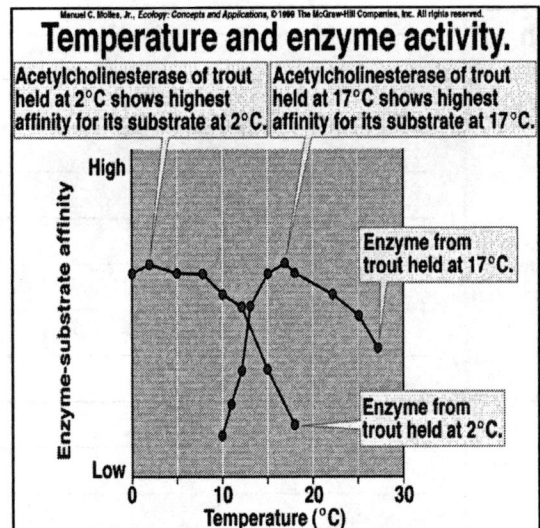

Figure 4.9

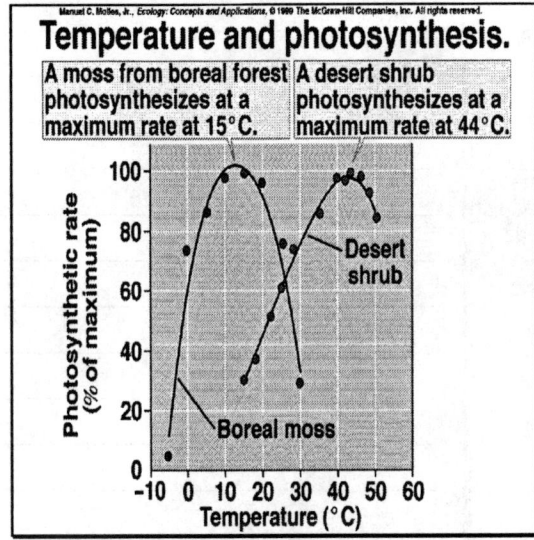

Figure 4.10

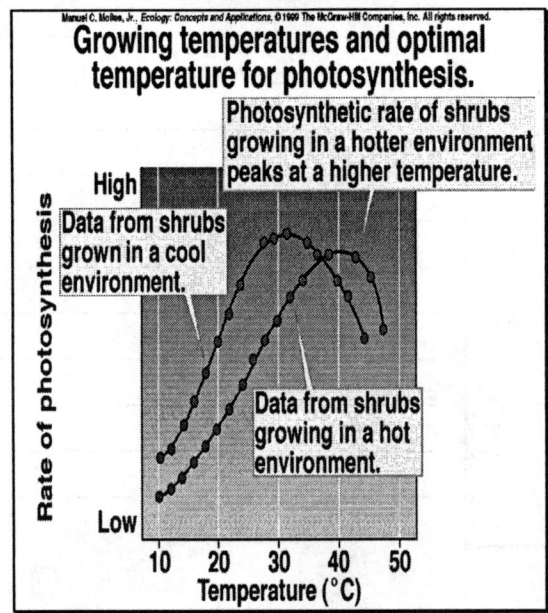

Figure 4.11

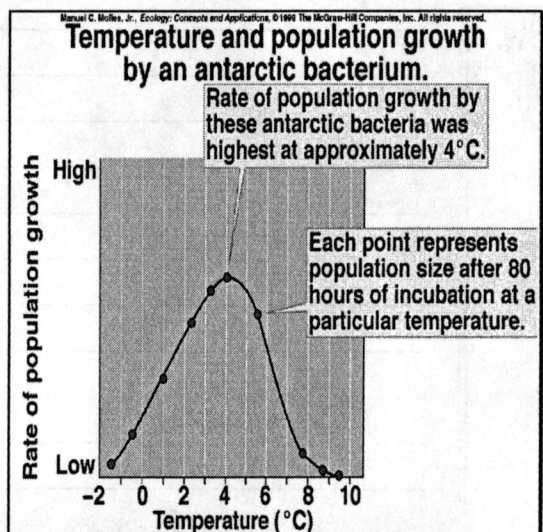

Figure 4.12

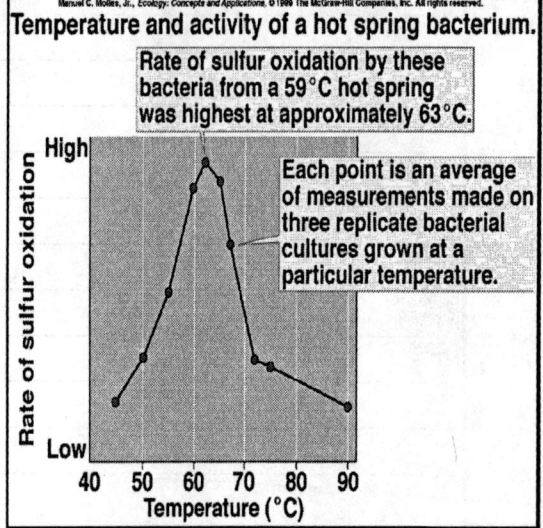

Figure 4.13

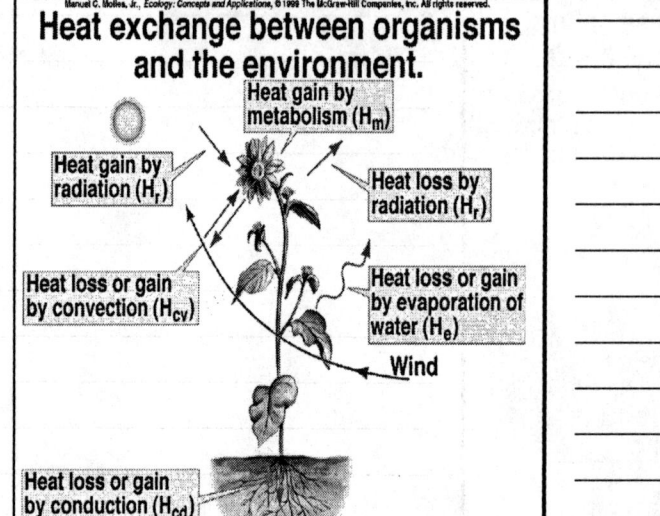

Figure 4.14

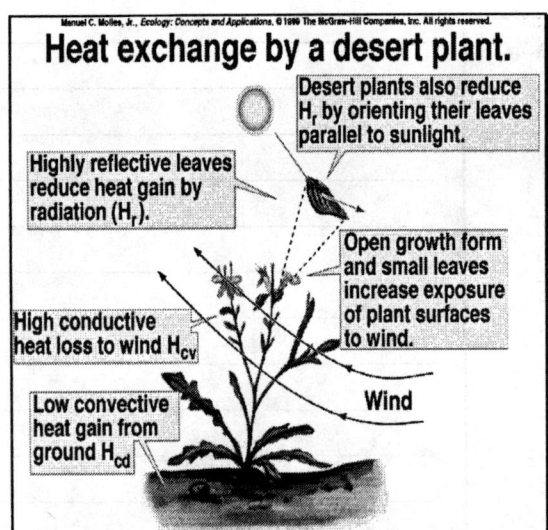

Figure 4.15

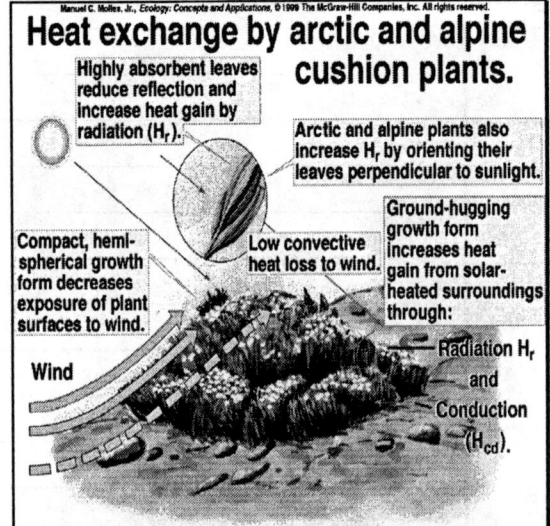

Figure 4.16

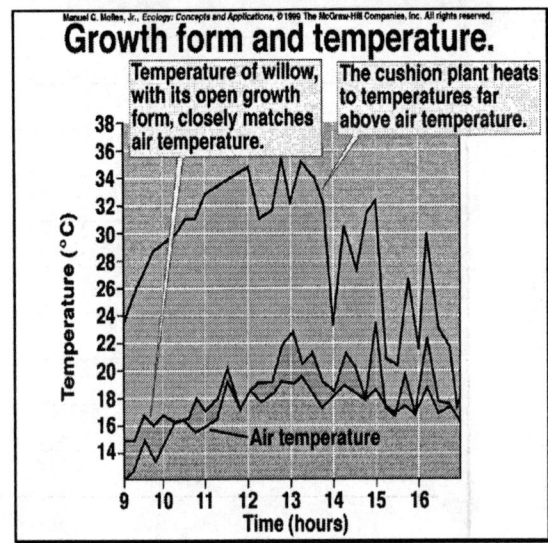

Figure 4.17

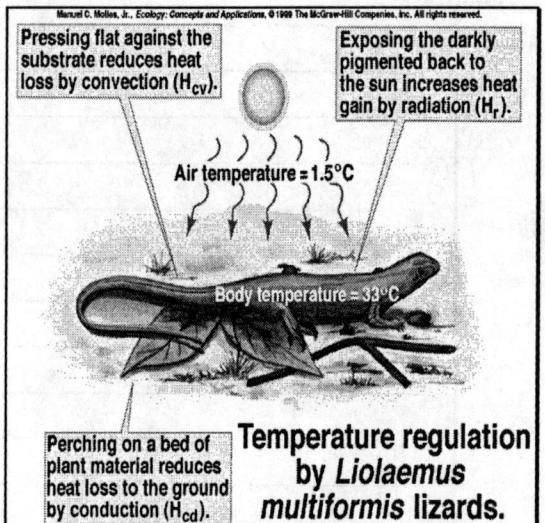

Figure 4.18

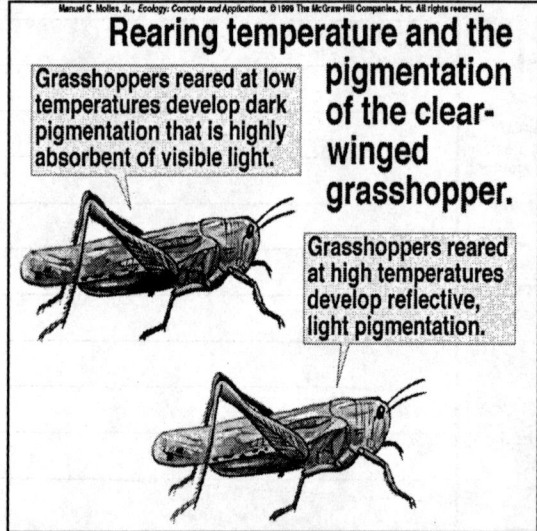

Figure 4.19

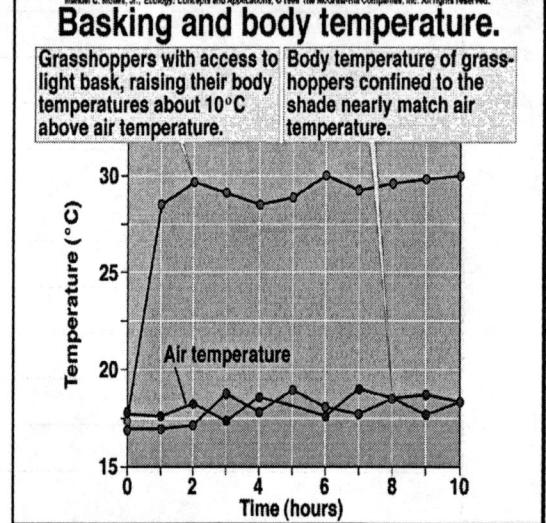

Figure 4.20

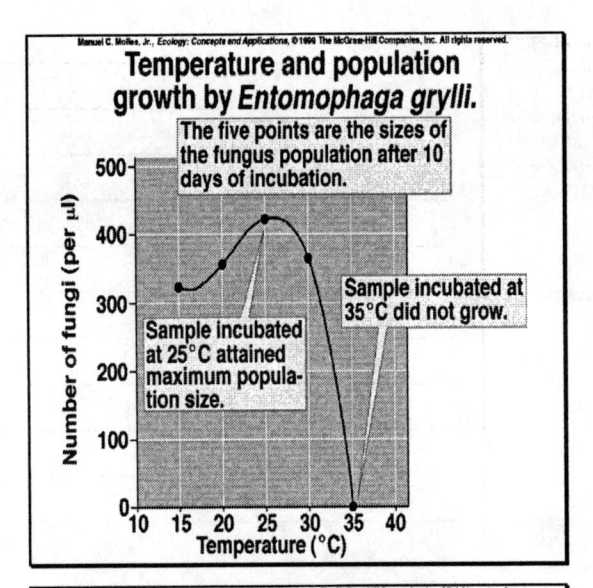

Temperature and population growth by *Entomophaga grylli*.

The five points are the sizes of the fungus population after 10 days of incubation.

Sample incubated at 25°C attained maximum population size.

Sample incubated at 35°C did not grow.

Figure 4.21

Temperature and metabolic rate of arctic and tropical mammals.

Tropical species maintain a constant metabolic rate over a narrow range of temperatures. (Sloth, Night monkey, Human, Marmoset)

Arctic species maintain a constant metabolic rate over a broad range of temperatures. (Ground squirrel, Polar bear club, Eskimo dog, Arctic fox)

Figure 4.22

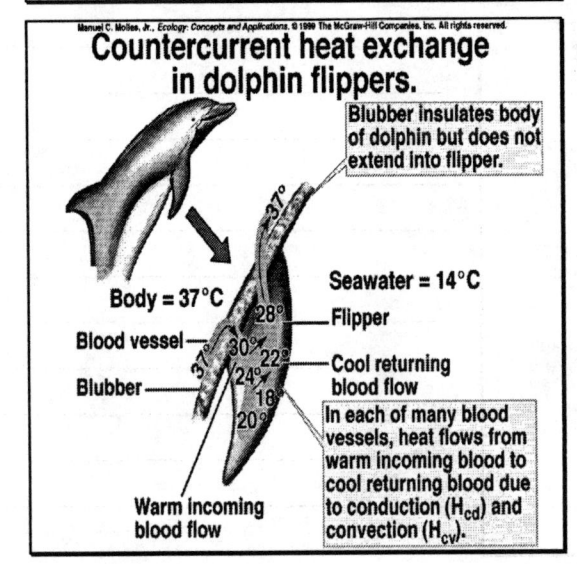

Countercurrent heat exchange in dolphin flippers.

Blubber insulates body of dolphin but does not extend into flipper.

Body = 37°C
Seawater = 14°C
Blood vessel
Blubber
Flipper
Cool returning blood flow
Warm incoming blood flow

In each of many blood vessels, heat flows from warm incoming blood to cool returning blood due to conduction (H_{cd}) and convection (H_{cv}).

Figure 4.23

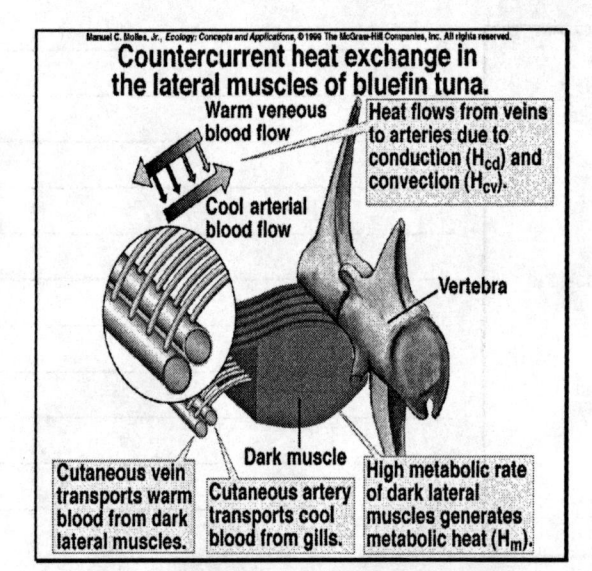

Figure 4.24

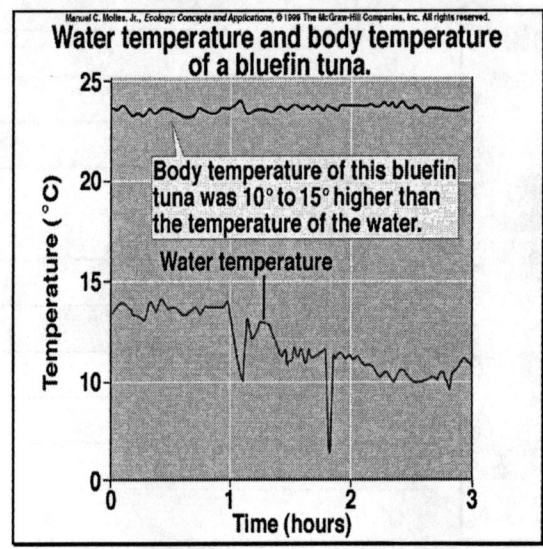

Figure 4.25 (1)

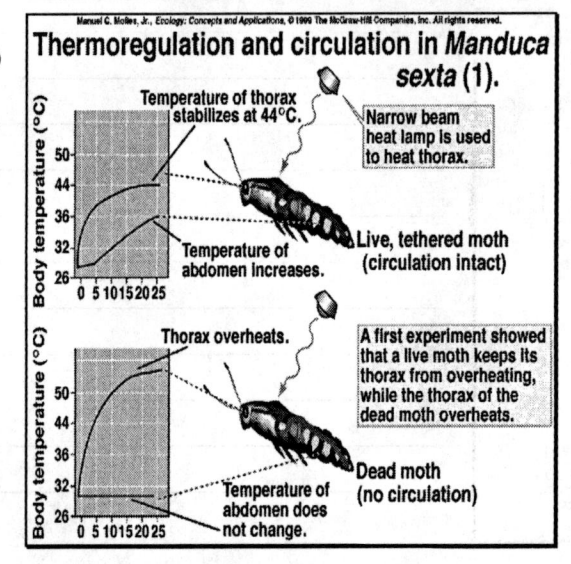

Figure 4.25 (2)

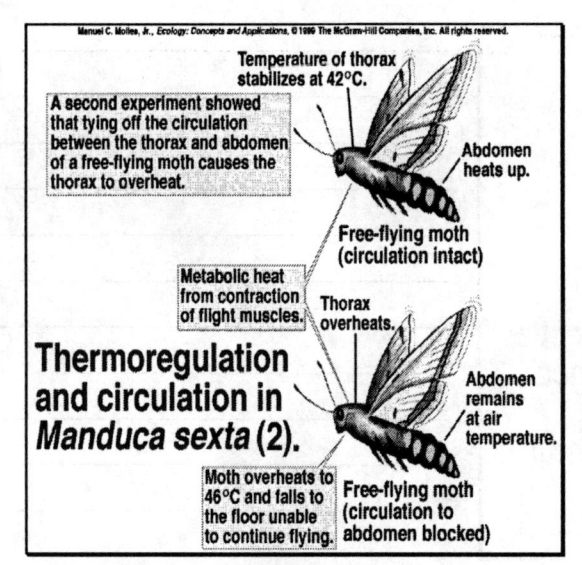

Thermoregulation and circulation in *Manduca sexta* (2).

- Temperature of thorax stabilizes at 42°C.
- A second experiment showed that tying off the circulation between the thorax and abdomen of a free-flying moth causes the thorax to overheat.
- Abdomen heats up.
- Free-flying moth (circulation intact)
- Metabolic heat from contraction of flight muscles.
- Thorax overheats.
- Abdomen remains at air temperature.
- Moth overheats to 46°C and falls to the floor unable to continue flying.
- Free-flying moth (circulation to abdomen blocked)

Figure 4.26

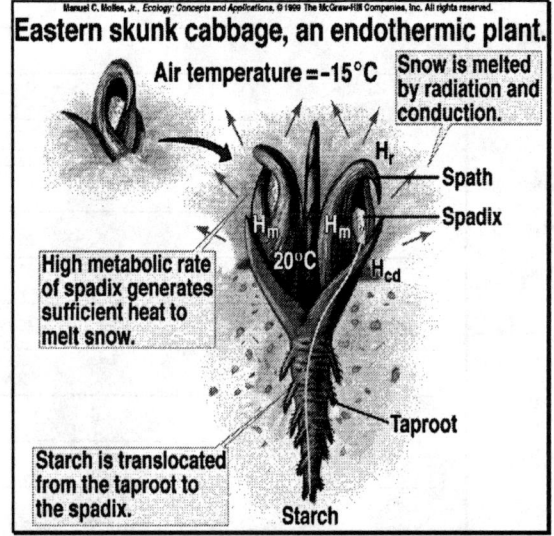

Eastern skunk cabbage, an endothermic plant.

- Air temperature = −15°C
- Snow is melted by radiation and conduction.
- H_r
- Spath
- Spadix
- H_m
- 20°C
- H_{cd}
- High metabolic rate of spadix generates sufficient heat to melt snow.
- Taproot
- Starch is translocated from the taproot to the spadix.
- Starch

Figure 4.27

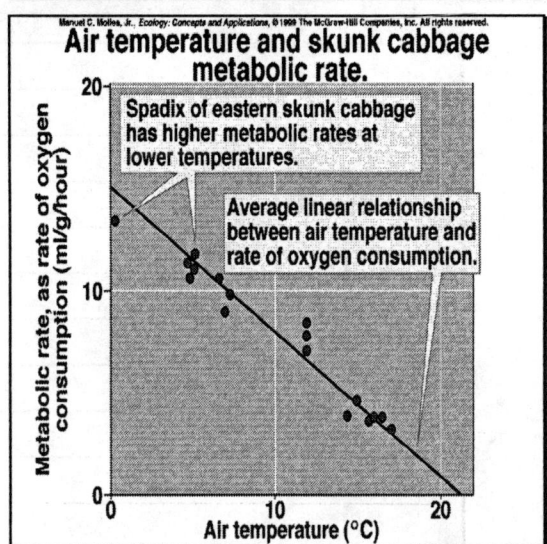

Air temperature and skunk cabbage metabolic rate.

- Spadix of eastern skunk cabbage has higher metabolic rates at lower temperatures.
- Average linear relationship between air temperature and rate of oxygen consumption.
- Metabolic rate, as rate of oxygen consumption (ml/g/hour)
- Air temperature (°C)

Figure 4.28

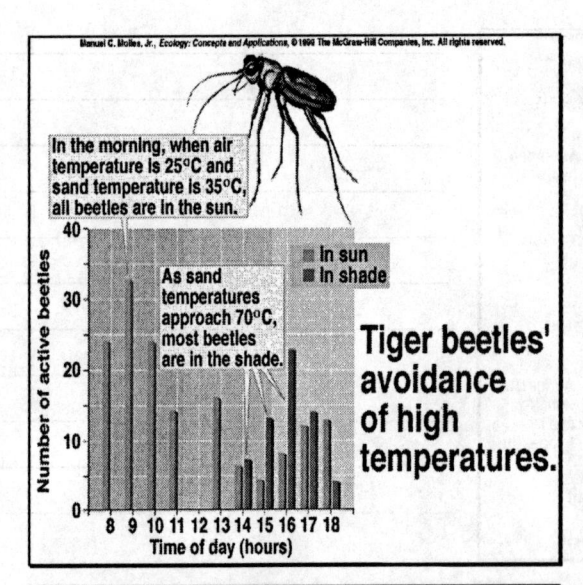

Tiger beetles' avoidance of high temperatures.

In the morning, when air temperature is 25°C and sand temperature is 35°C, all beetles are in the sun.

As sand temperatures approach 70°C, most beetles are in the shade.

Figure 4.29

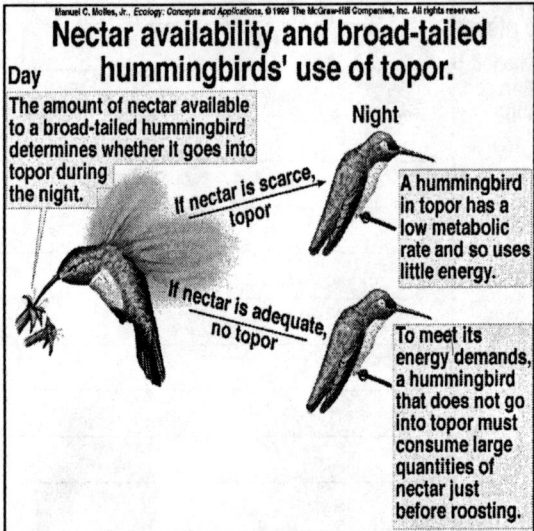

Nectar availability and broad-tailed hummingbirds' use of topor.

The amount of nectar available to a broad-tailed hummingbird determines whether it goes into topor during the night.

If nectar is scarce, topor

If nectar is adequate, no topor

A hummingbird in topor has a low metabolic rate and so uses little energy.

To meet its energy demands, a hummingbird that does not go into topor must consume large quantities of nectar just before roosting.

Figure 4.30

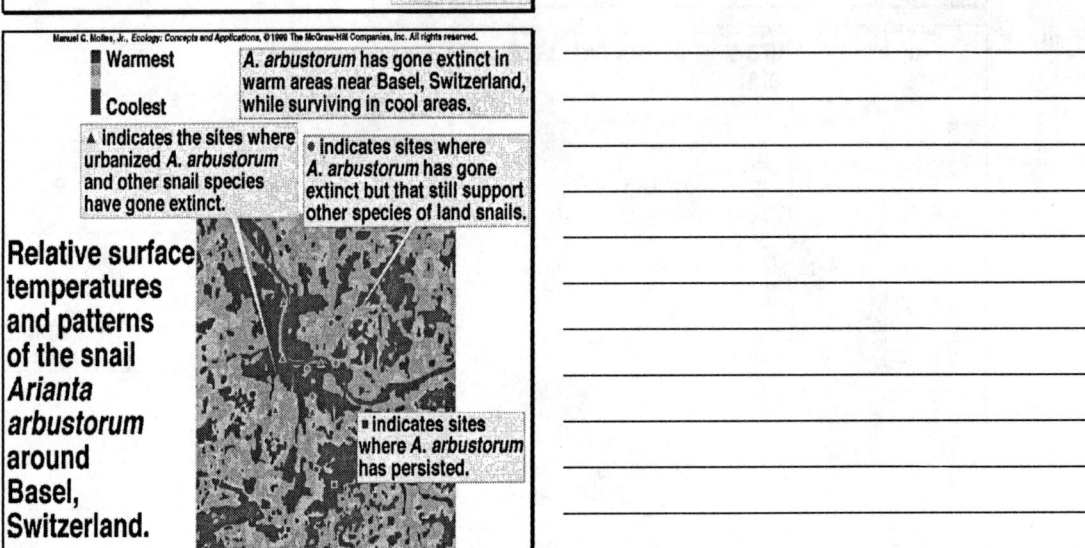

Relative surface temperatures and patterns of the snail *Arianta arbustorum* around Basel, Switzerland.

■ Warmest
■ Coolest

A. arbustorum has gone extinct in warm areas near Basel, Switzerland, while surviving in cool areas.

▲ indicates the sites where urbanized *A. arbustorum* and other snail species have gone extinct.

● indicates sites where *A. arbustorum* has gone extinct but that still support other species of land snails.

■ indicates sites where *A. arbustorum* has persisted.

Figure 4.31

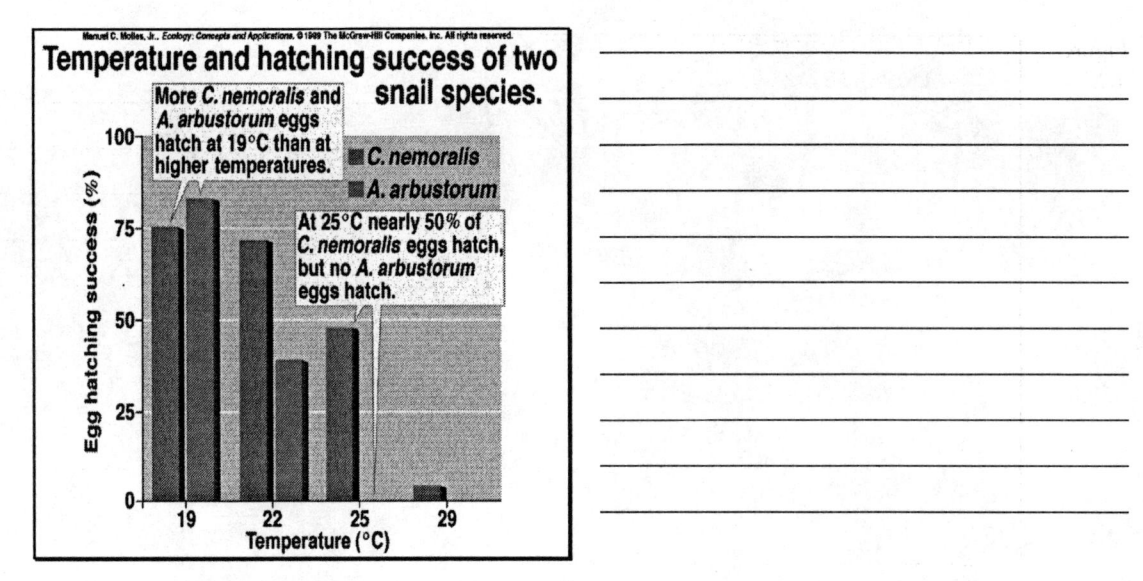

Temperature and hatching success of two snail species.

More *C. nemoralis* and *A. arbustorum* eggs hatch at 19°C than at higher temperatures.

At 25°C nearly 50% of *C. nemoralis* eggs hatch, but no *A. arbustorum* eggs hatch.

Figure 5.1

Chapter 5

Ecology

Concepts and Applications
First Edition

Manuel C. Molles, Jr.

(c) The McGraw-Hill Companies, Inc.

Figure 5.1

Air temperature of 46°C is higher than lethal maximum for the cicada.

Falling to the ground, with a temperature of 70°C, would be certain death for the cicada.

An ecological puzzle.

How does the cicada remain active when environmental temperatures exceed its lethal maximum?

Figure 5.2

Water vapor in air can be measured either as grams of vapor per cubic meter of air or by the pressure exerted by the water vapor in air.

Air temperature and two measures of water vapor saturation of air.

At low temperatures, air is saturated by low quantities of water vapor and water vapor pressure is low.

As temperature increases, the amount of water air holds at saturation and saturation water vapor pressure increases.

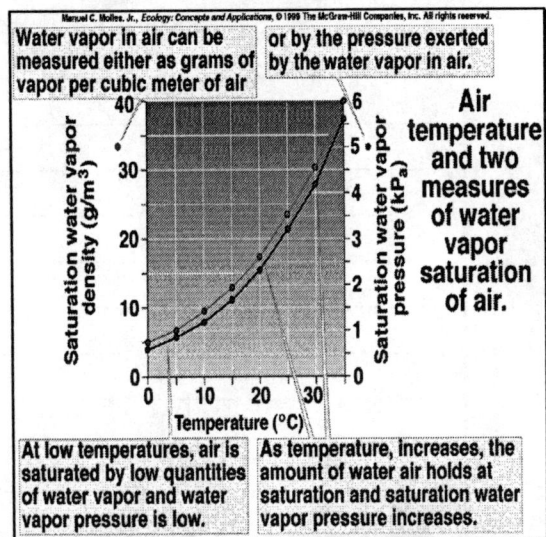

Figure 5.3

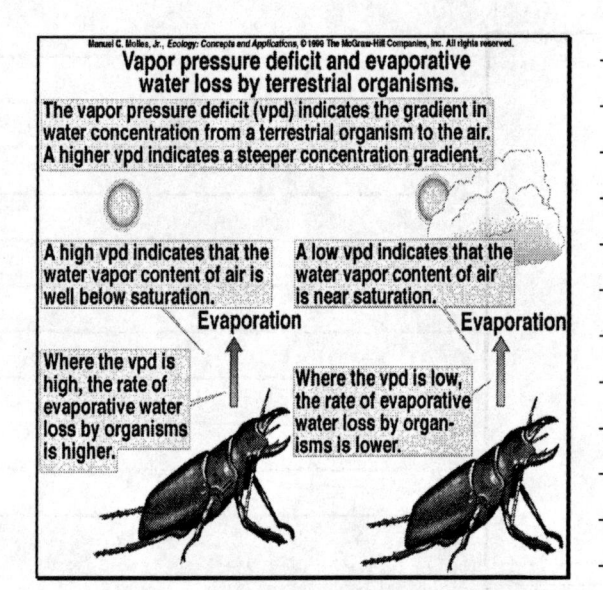

Vapor pressure deficit and evaporative water loss by terrestrial organisms.

The vapor pressure deficit (vpd) indicates the gradient in water concentration from a terrestrial organism to the air. A higher vpd indicates a steeper concentration gradient.

A high vpd indicates that the water vapor content of air is well below saturation.

A low vpd indicates that the water vapor content of air is near saturation.

Where the vpd is high, the rate of evaporative water loss by organisms is higher.

Where the vpd is low, the rate of evaporative water loss by organisms is lower.

Figure 5.4

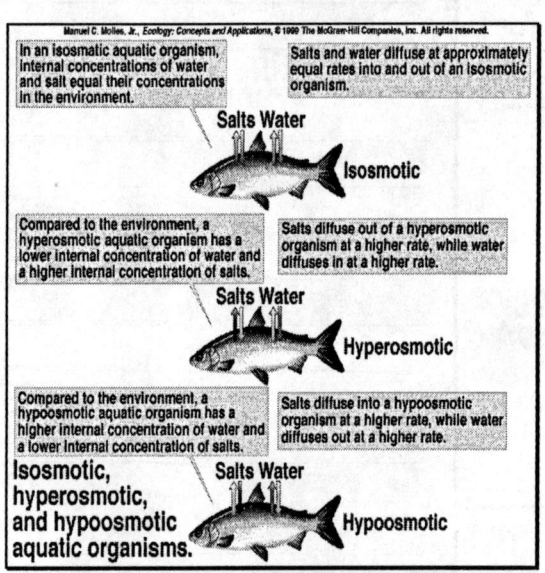

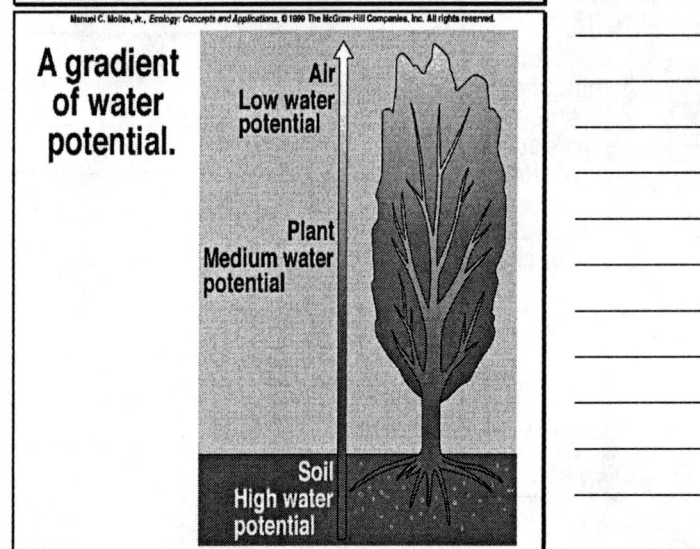

In an isosmotic aquatic organism, internal concentrations of water and salt equal their concentrations in the environment.

Salts and water diffuse at approximately equal rates into and out of an isosmotic organism.

Isosmotic

Compared to the environment, a hyperosmotic aquatic organism has a lower internal concentration of water and a higher internal concentration of salts.

Salts diffuse out of a hyperosmotic organism at a higher rate, while water diffuses in at a higher rate.

Hyperosmotic

Compared to the environment, a hypoosmotic aquatic organism has a higher internal concentration of water and a lower internal concentration of salts.

Salts diffuse into a hypoosmotic organism at a higher rate, while water diffuses out at a higher rate.

Isosmotic, hyperosmotic, and hypoosmotic aquatic organisms.

Hypoosmotic

Figure 5.5

A gradient of water potential.

Air — Low water potential

Plant — Medium water potential

Soil — High water potential

Figure 5.6

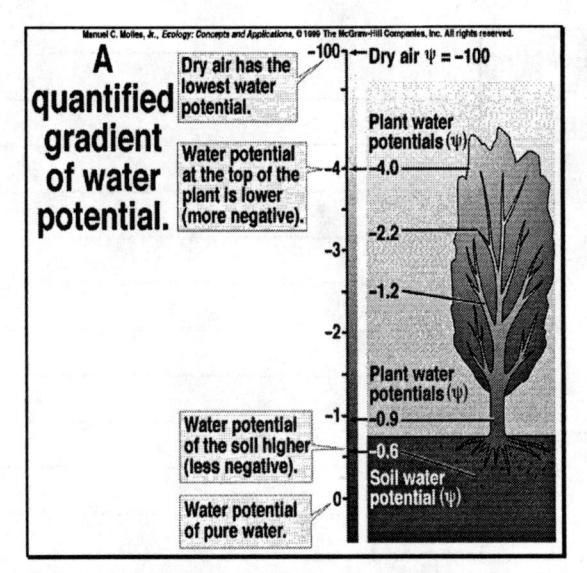

Figure 5.7

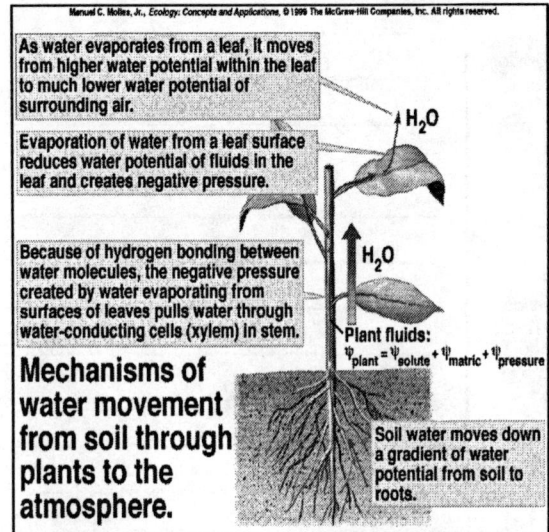

Figure 5.8

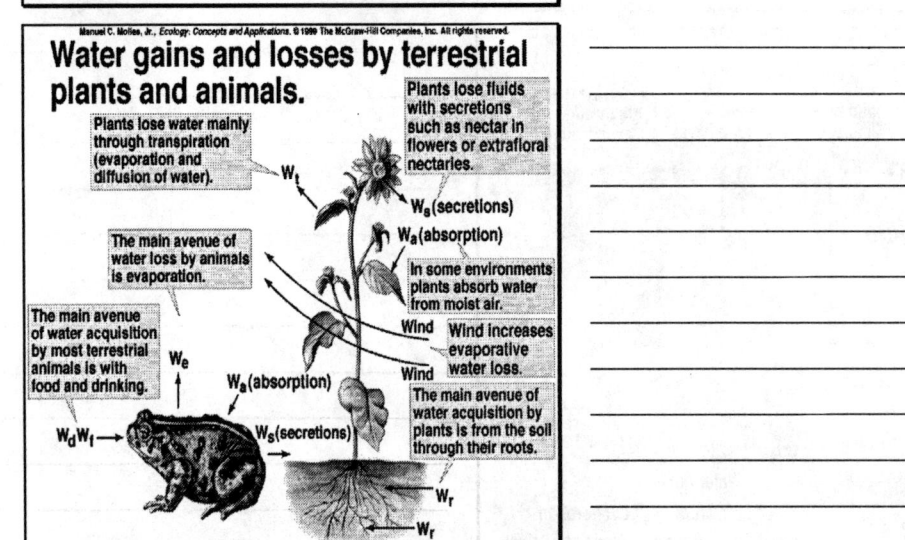

Figure 5.9

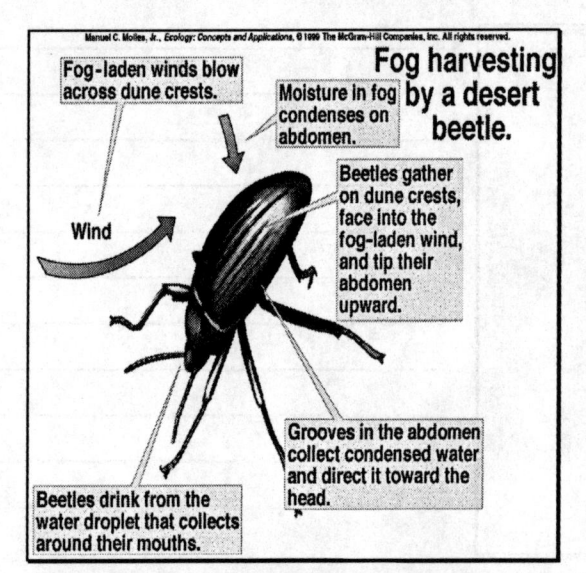

Fog harvesting by a desert beetle.

- Fog-laden winds blow across dune crests.
- Moisture in fog condenses on abdomen.
- Wind
- Beetles gather on dune crests, face into the fog-laden wind, and tip their abdomen upward.
- Grooves in the abdomen collect condensed water and direct it toward the head.
- Beetles drink from the water droplet that collects around their mouths.

Figure 5.10

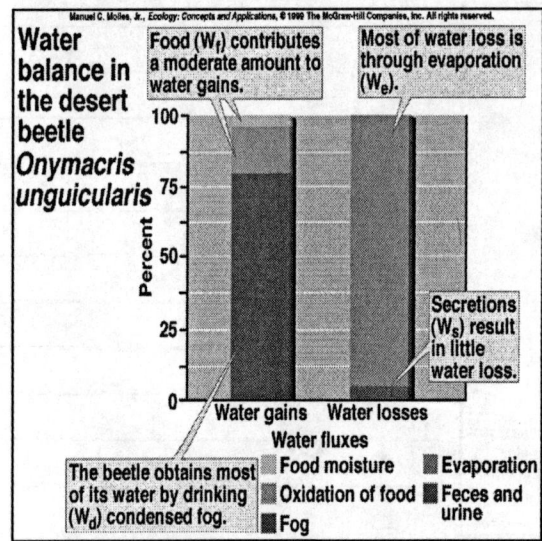

Water balance in the desert beetle *Onymacris unguicularis*

- Food (W_f) contributes a moderate amount to water gains.
- Most of water loss is through evaporation (W_e).
- Secretions (W_s) result in little water loss.
- The beetle obtains most of its water by drinking (W_d) condensed fog.

Legend: Food moisture, Oxidation of food, Fog, Evaporation, Feces and urine

Figure 5.11

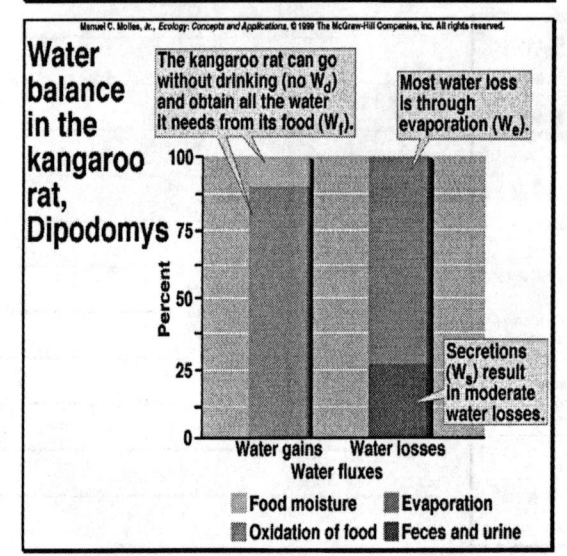

Water balance in the kangaroo rat, *Dipodomys*

- The kangaroo rat can go without drinking (no W_d) and obtain all the water it needs from its food (W_f).
- Most water loss is through evaporation (W_e).
- Secretions (W_s) result in moderate water losses.

Legend: Food moisture, Oxidation of food, Evaporation, Feces and urine

Figure 5.12

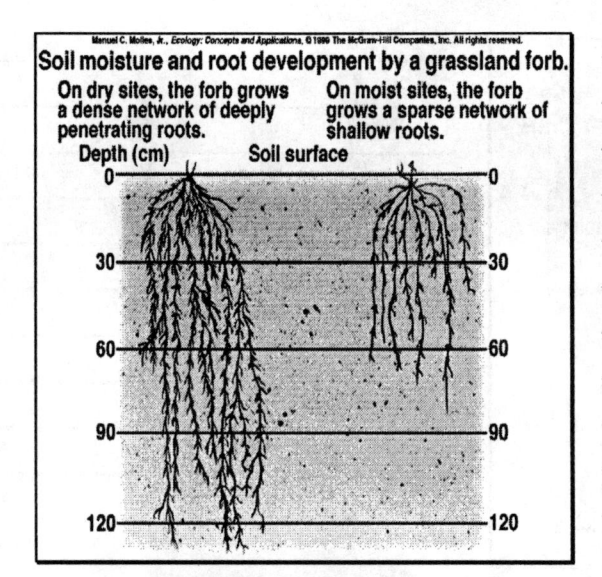

Soil moisture and root development by a grassland forb.

On dry sites, the forb grows a dense network of deeply penetrating roots.

On moist sites, the forb grows a sparse network of shallow roots.

Figure 5.13

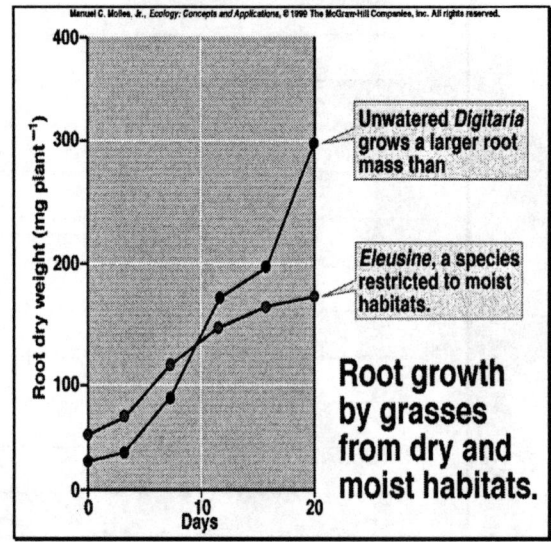

Unwatered *Digitaria* grows a larger root mass than

Eleusine, a species restricted to moist habitats.

Root growth by grasses from dry and moist habitats.

Figure 5.14

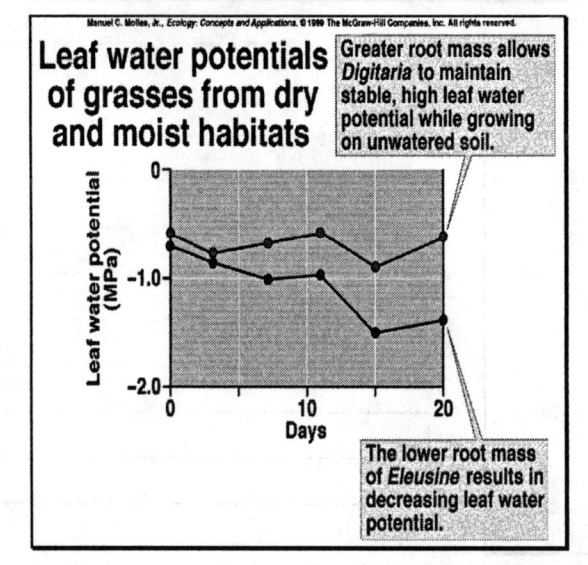

Leaf water potentials of grasses from dry and moist habitats

Greater root mass allows *Digitaria* to maintain stable, high leaf water potential while growing on unwatered soil.

The lower root mass of *Eleusine* results in decreasing leaf water potential.

Figure 5.15

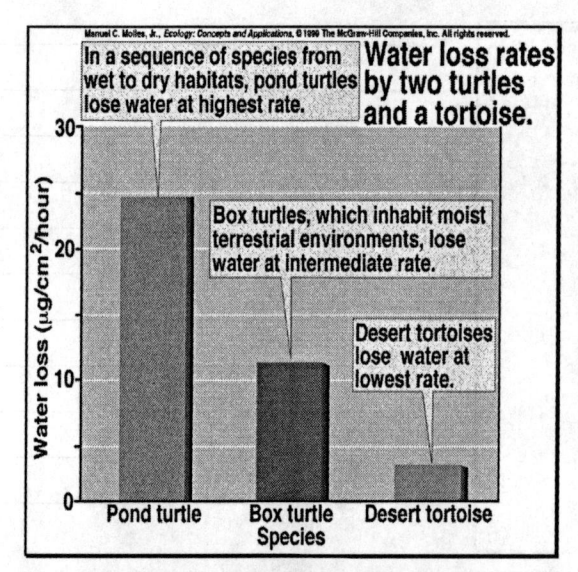

Water loss rates by two turtles and a tortoise.

In a sequence of species from wet to dry habitats, pond turtles lose water at highest rate.

Box turtles, which inhabit moist terrestrial environments, lose water at intermediate rate.

Desert tortoises lose water at lowest rate.

Water loss ($\mu g/cm^2/hour$) vs Species: Pond turtle, Box turtle, Desert tortoise

Figure 5.16

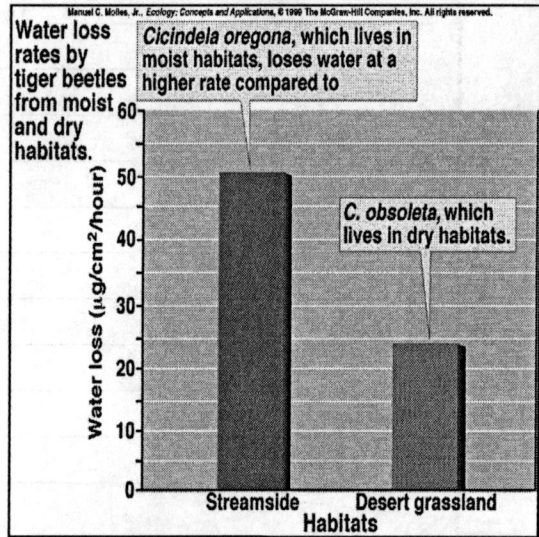

Water loss rates by tiger beetles from moist and dry habitats.

Cicindela oregona, which lives in moist habitats, loses water at a higher rate compared to *C. obsoleta*, which lives in dry habitats.

Water loss ($\mu g/cm^2/hour$) vs Habitats: Streamside, Desert grassland

Figure 5.17

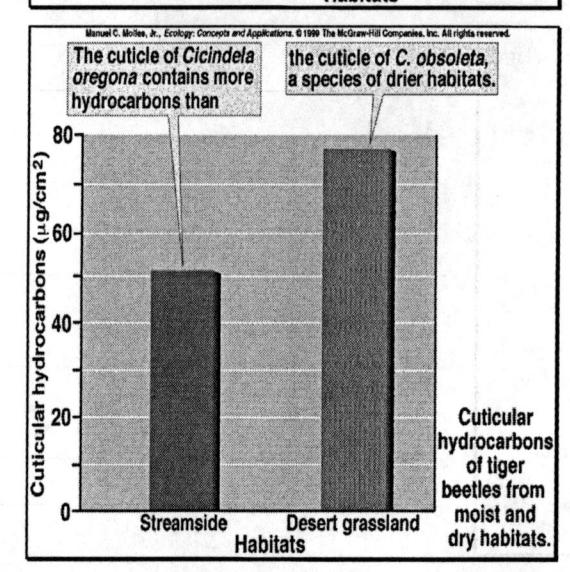

Cuticular hydrocarbons of tiger beetles from moist and dry habitats.

The cuticle of *Cicindela oregona* contains more hydrocarbons than the cuticle of *C. obsoleta*, a species of drier habitats.

Cuticular hydrocarbons ($\mu g/cm^2$) vs Habitats: Streamside, Desert grassland

Figure 5.19

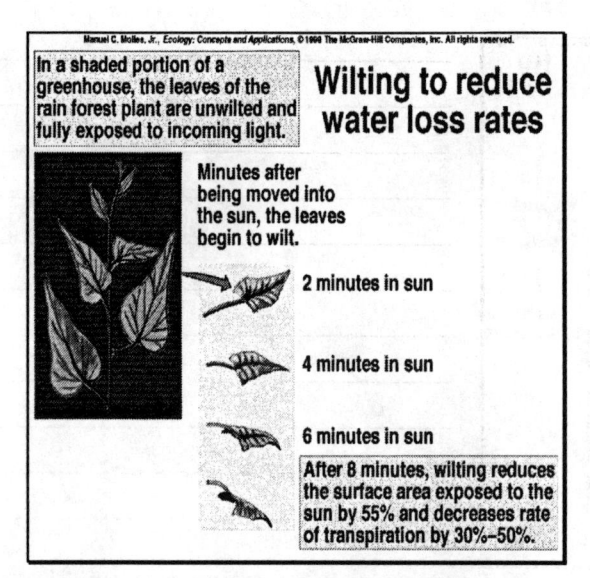

Wilting to reduce water loss rates

In a shaded portion of a greenhouse, the leaves of the rain forest plant are unwilted and fully exposed to incoming light.

Minutes after being moved into the sun, the leaves begin to wilt.

2 minutes in sun

4 minutes in sun

6 minutes in sun

After 8 minutes, wilting reduces the surface area exposed to the sun by 55% and decreases rate of transpiration by 30%–50%.

Figure 5.21

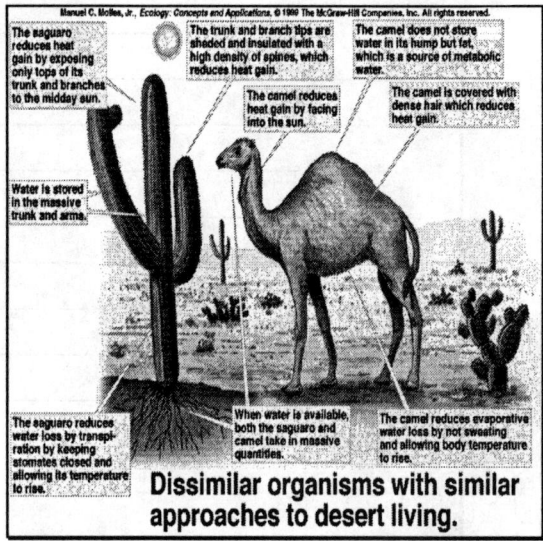

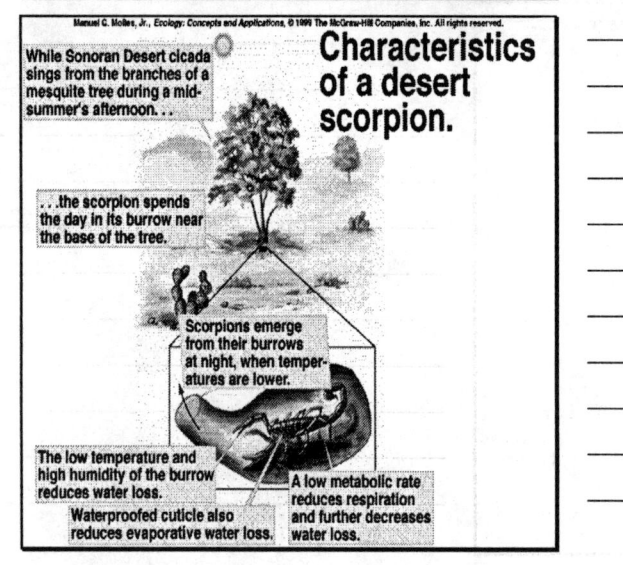

The saguaro reduces heat gain by exposing only tops of its trunk and branches to the midday sun.

The trunk and branch tips are shaded and insulated with a high density of spines, which reduces heat gain.

The camel does not store water in its hump but fat, which is a source of metabolic water.

The camel reduces heat gain by facing into the sun.

The camel is covered with dense hair which reduces heat gain.

Water is stored in the massive trunk and arms.

The saguaro reduces water loss by transpiration by keeping stomates closed and allowing its temperature to rise.

When water is available, both the saguaro and camel take in massive quantities.

The camel reduces evaporative water loss by not sweating and allowing body temperature to rise.

Dissimilar organisms with similar approaches to desert living.

Figure 5.22

Characteristics of a desert scorpion.

While Sonoran Desert cicada sings from the branches of a mesquite tree during a midsummer's afternoon...

...the scorpion spends the day in its burrow near the base of the tree.

Scorpions emerge from their burrows at night, when temperatures are lower.

The low temperature and high humidity of the burrow reduces water loss.

Waterproofed cuticle also reduces evaporative water loss.

A low metabolic rate reduces respiration and further decreases water loss.

Figure 5.23

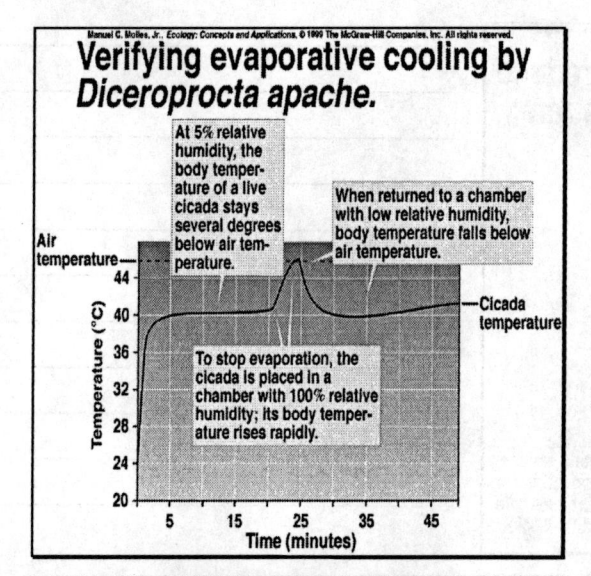

Verifying evaporative cooling by *Diceroprocta apache*.

At 5% relative humidity, the body temperature of a live cicada stays several degrees below air temperature.

When returned to a chamber with low relative humidity, body temperature falls below air temperature.

To stop evaporation, the cicada is placed in a chamber with 100% relative humidity; its body temperature rises rapidly.

Air temperature — Cicada temperature

Temperature (°C) vs Time (minutes)

Figure 5.24

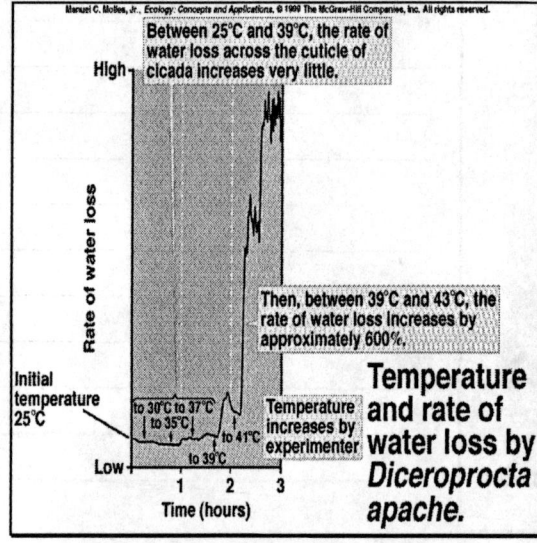

Temperature and rate of water loss by *Diceroprocta apache*.

Between 25°C and 39°C, the rate of water loss across the cuticle of cicada increases very little.

Then, between 39°C and 43°C, the rate of water loss increases by approximately 600%.

Initial temperature 25°C. Temperature increases by experimenter: to 30°C, to 35°C, to 37°C, to 39°C, to 41°C.

Rate of water loss vs Time (hours)

Figure 5.26

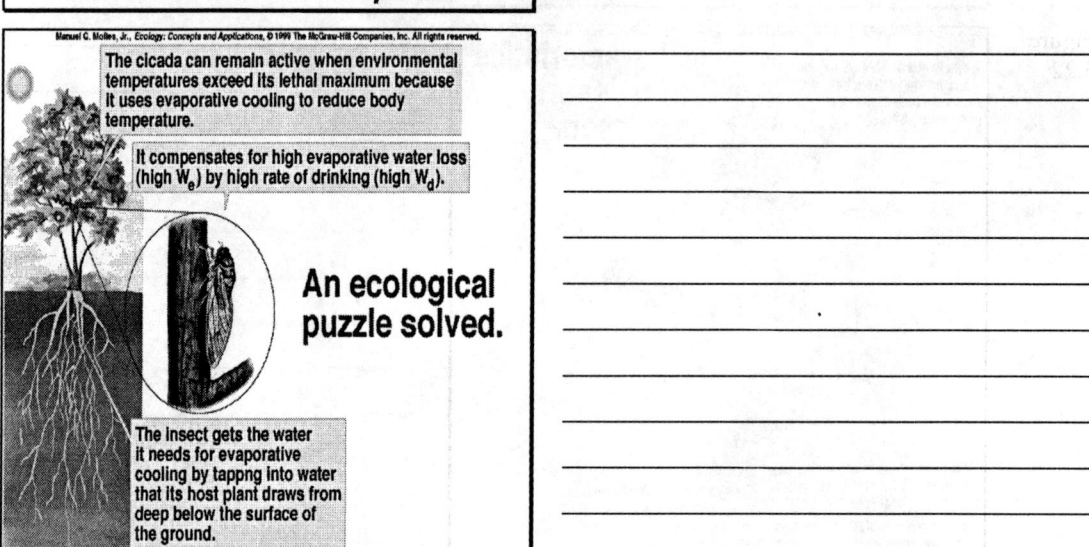

An ecological puzzle solved.

The cicada can remain active when environmental temperatures exceed its lethal maximum because it uses evaporative cooling to reduce body temperature.

It compensates for high evaporative water loss (high W_e) by high rate of drinking (high W_d).

The insect gets the water it needs for evaporative cooling by tapping into water that its host plant draws from deep below the surface of the ground.

Figure 5.27

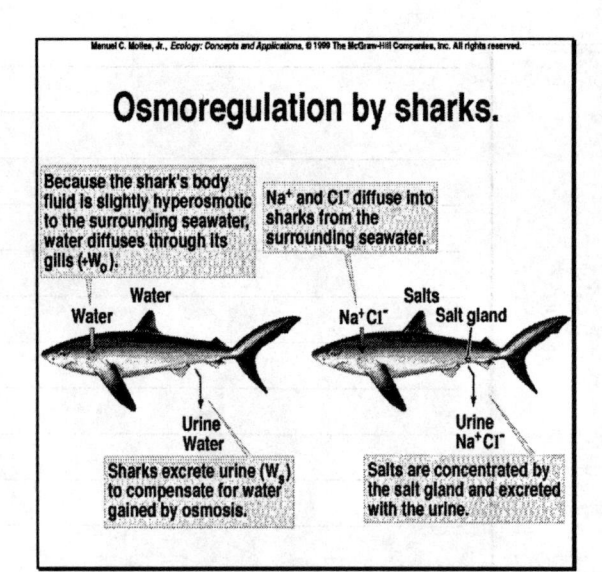

Figure 5.28

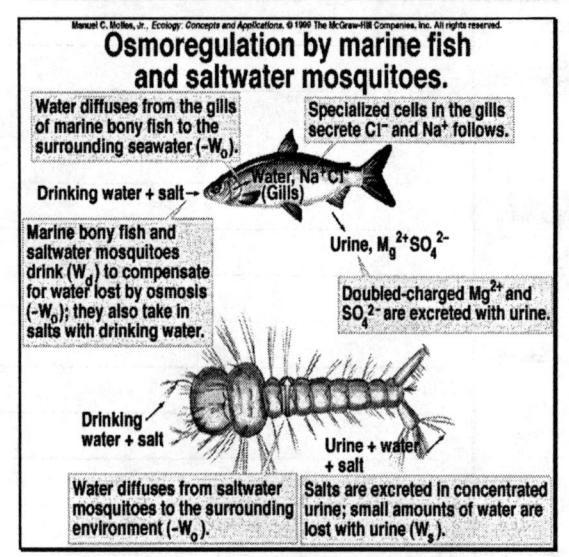

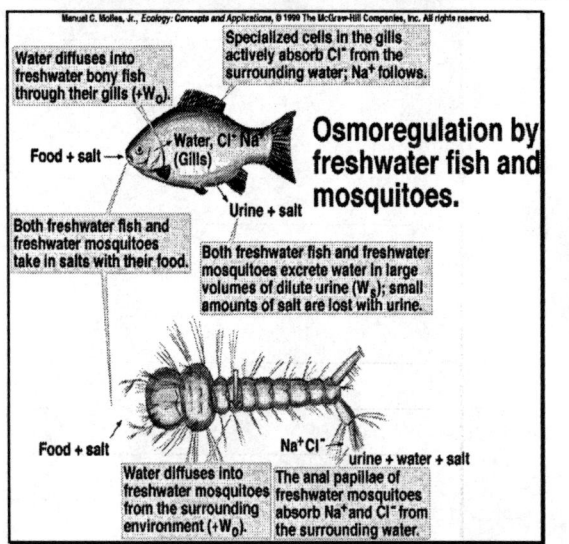

Figure 5.29

Figure 5.30

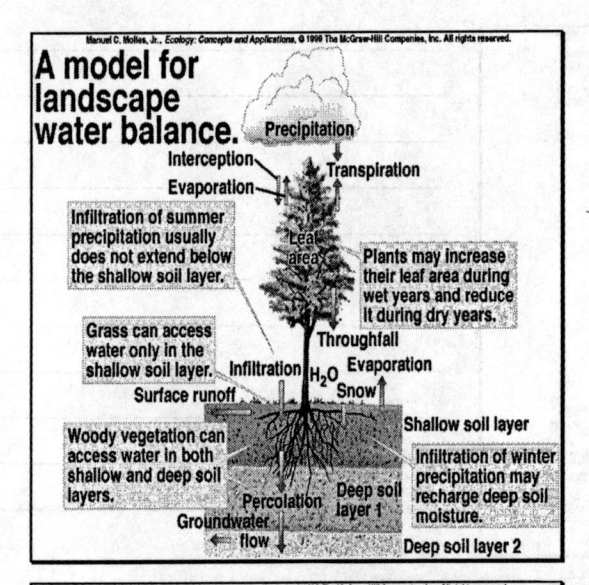

Figure 5.31

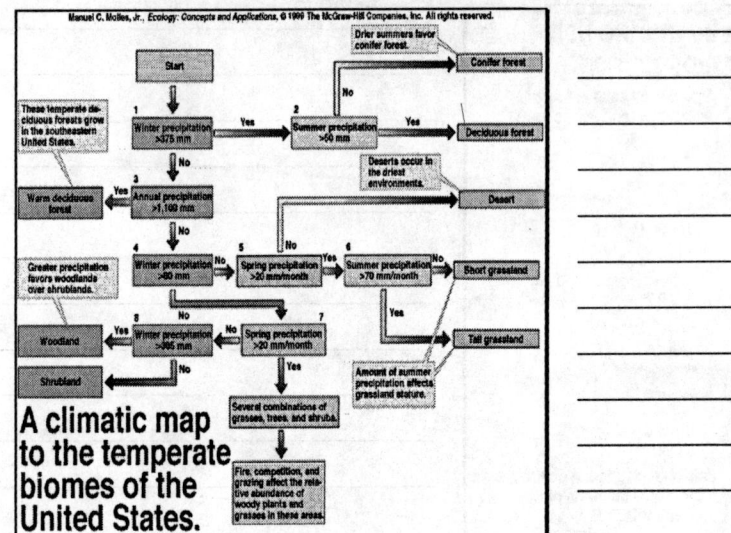

Chapter 6

Ecology

Concepts and Applications
First Edition

Manuel C. Molles, Jr.

(c) The McGraw-Hill Companies, Inc.

Figure 6.2: Trophic diversity across the biological kingdoms.

- Bacteria draw on a greater variety of energy sources than any other group of organisms.
- Protists include many heterotrophic and photosynthetic species.
- Plants are mainly photosynthetic, with a few heterotrophic species.
- Fungi and animals are all heterotrophic.

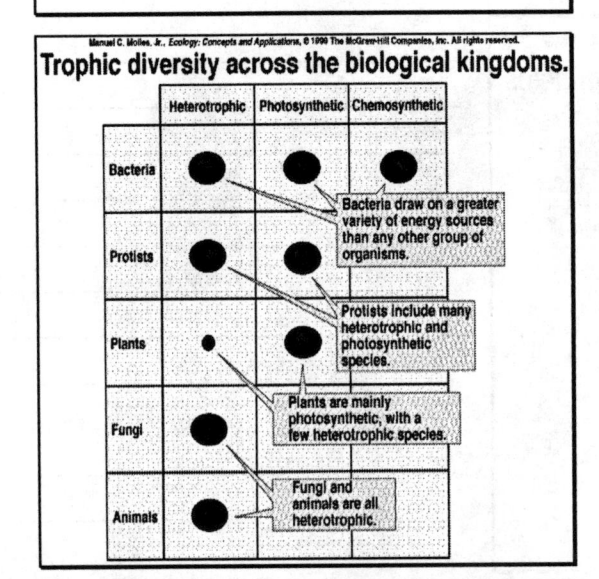

Figure 6.3: Photosynthetically active radiation (PAR) in a boreal forest.

- 100%
- Boreal forests reflect about 10% of incoming PAR.
- The canopy absorbs 79% of PAR.
- Plants in the middle layers absorb an additional 7% of PAR.
- Low vegetation absorbs about 2% of PAR.
- Only about 2% of PAR shining on the canopy reaches the forest floor.

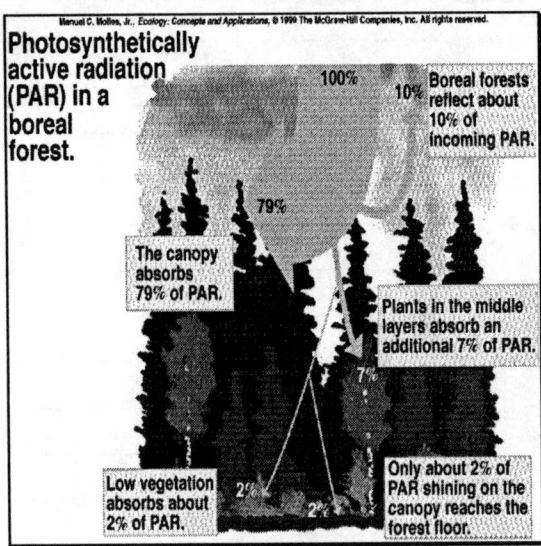

Figure 6.4

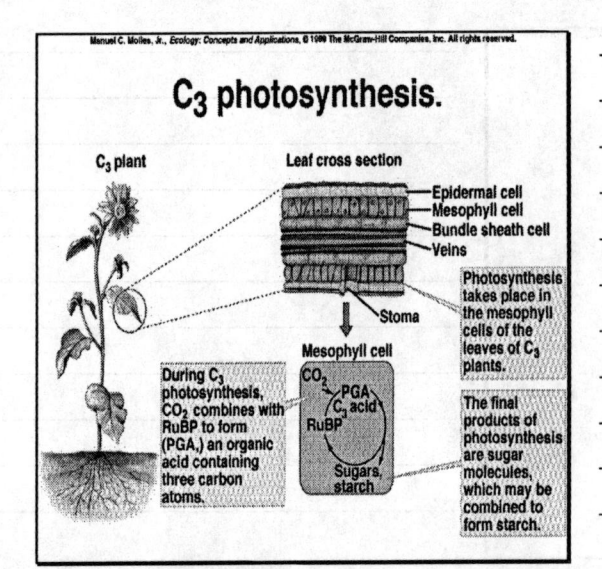

Figure 6.5

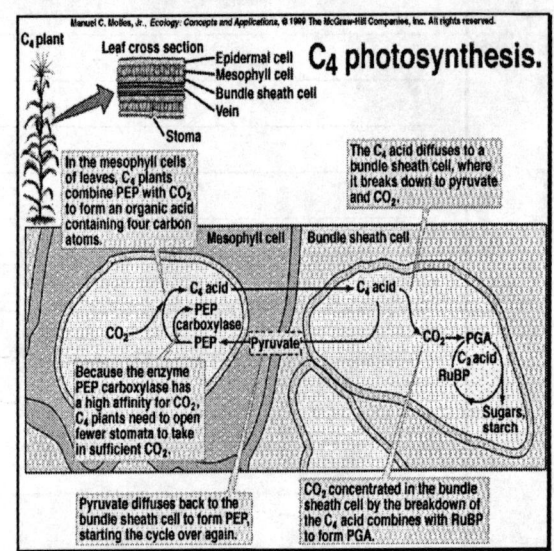

Figure 6.6

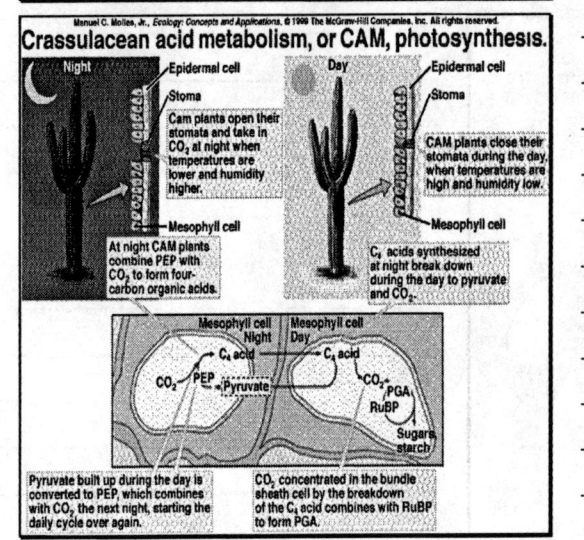

Figure 6.7

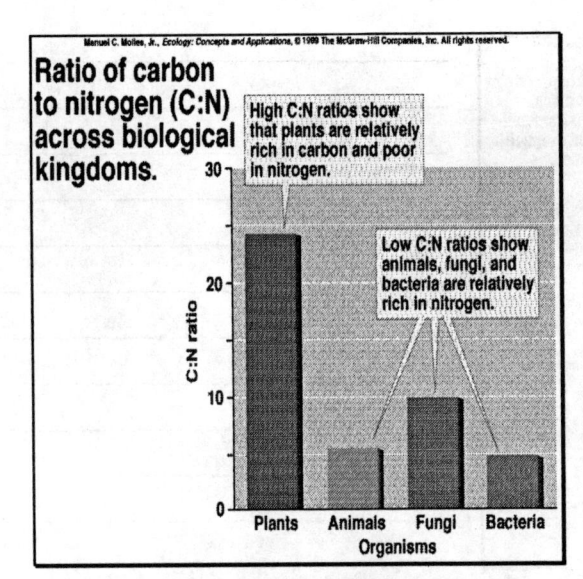

Ratio of carbon to nitrogen (C:N) across biological kingdoms.

High C:N ratios show that plants are relatively rich in carbon and poor in nitrogen.

Low C:N ratios show animals, fungi, and bacteria are relatively rich in nitrogen.

Organisms: Plants, Animals, Fungi, Bacteria

Figure 6.9

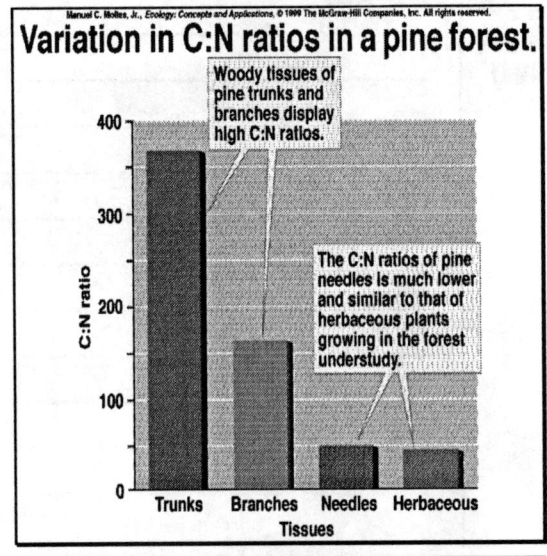

Variation in C:N ratios in a pine forest.

Woody tissues of pine trunks and branches display high C:N ratios.

The C:N ratios of pine needles is much lower and similar to that of herbaceous plants growing in the forest understudy.

Tissues: Trunks, Branches, Needles, Herbaceous

Figure 6.10

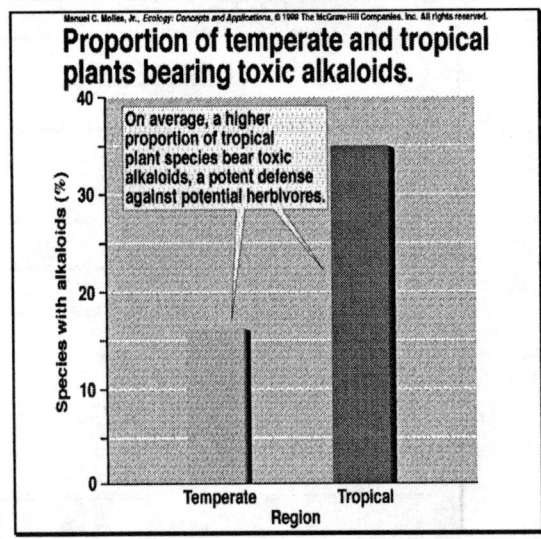

Proportion of temperate and tropical plants bearing toxic alkaloids.

On average, a higher proportion of tropical plant species bear toxic alkaloids, a potent defense against potential herbivores.

Region: Temperate, Tropical

Figure 6.11

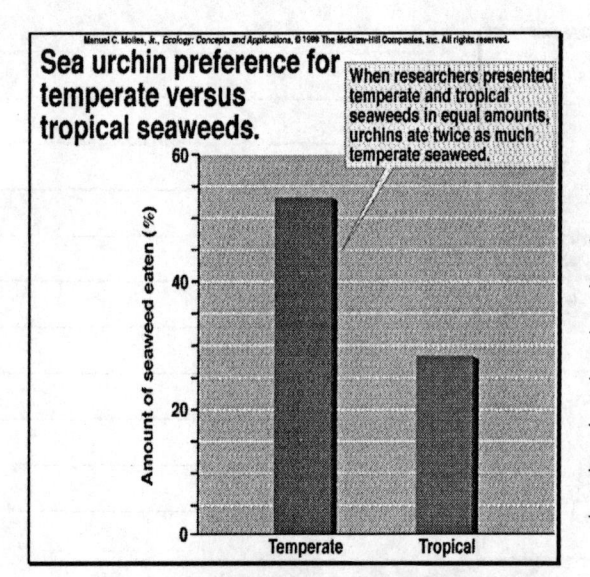

Sea urchin preference for temperate versus tropical seaweeds.

When researchers presented temperate and tropical seaweeds in equal amounts, urchins ate twice as much temperate seaweed.

Figure 6.12

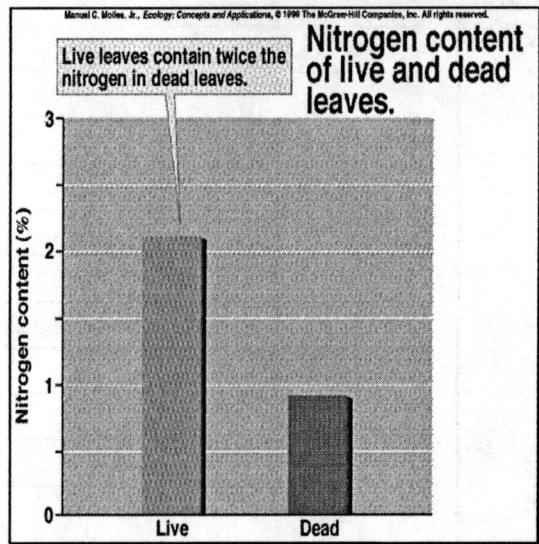

Nitrogen content of live and dead leaves.

Live leaves contain twice the nitrogen in dead leaves.

Figure 6.14

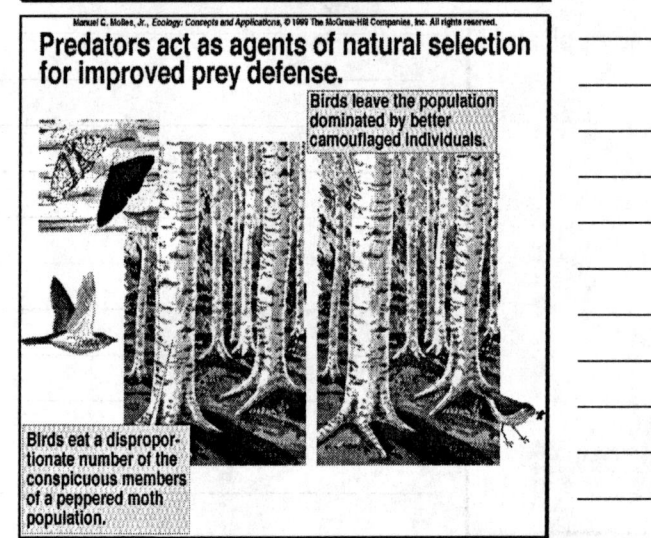

Predators act as agents of natural selection for improved prey defense.

Birds leave the population dominated by better camouflaged individuals.

Birds eat a disproportionate number of the conspicuous members of a peppered moth population.

Figure 6.15

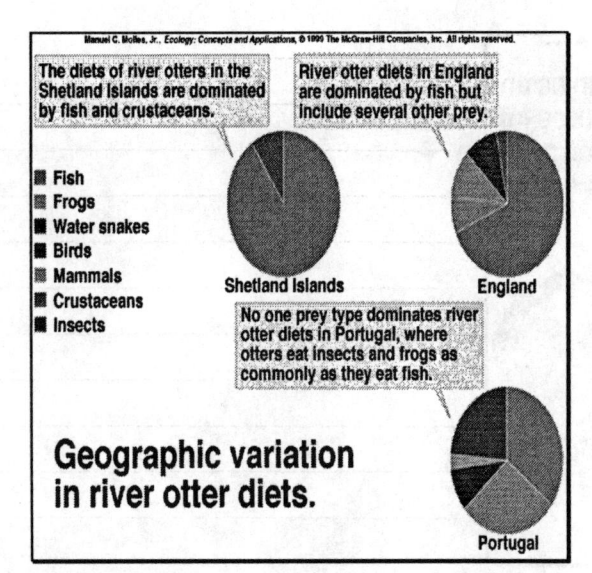

Figure 6.16

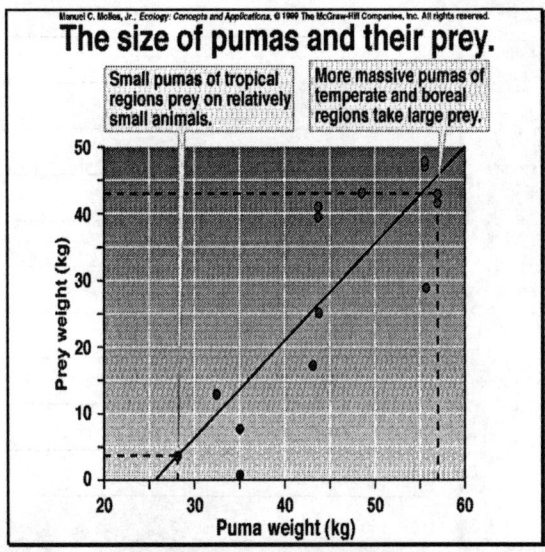

Figure 6.17

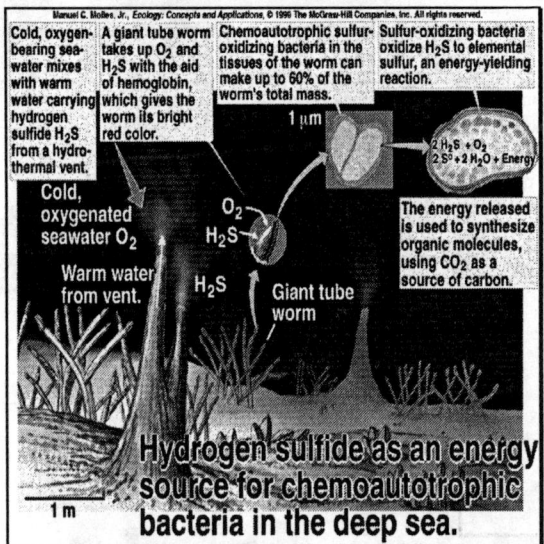

Figure 6.18

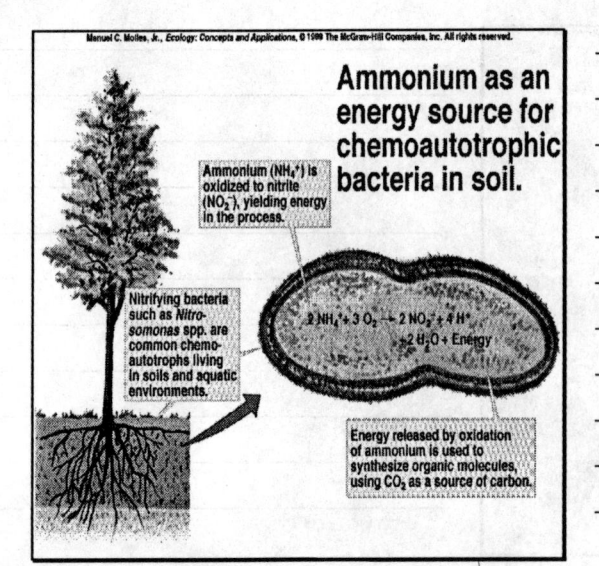

Ammonium as an energy source for chemoautotrophic bacteria in soil.

Ammonium (NH_4^+) is oxidized to nitrite (NO_2^-), yielding energy in the process.

Nitrifying bacteria such as *Nitrosomonas* spp. are common chemoautotrophs living in soils and aquatic environments.

$2 NH_4^+ + 3 O_2 \rightarrow 2 NO_2^- + 4 H^+ + 2 H_2O + Energy$

Energy released by oxidation of ammonium is used to synthesize organic molecules, using CO_2 as a source of carbon.

Figure 6.19

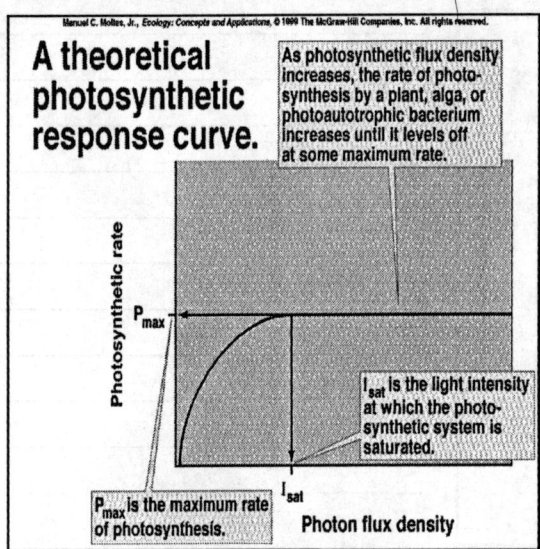

A theoretical photosynthetic response curve.

As photosynthetic flux density increases, the rate of photosynthesis by a plant, alga, or photoautotrophic bacterium increases until it levels off at some maximum rate.

I_{sat} is the light intensity at which the photosynthetic system is saturated.

P_{max} is the maximum rate of photosynthesis.

Photon flux density

Figure 6.20

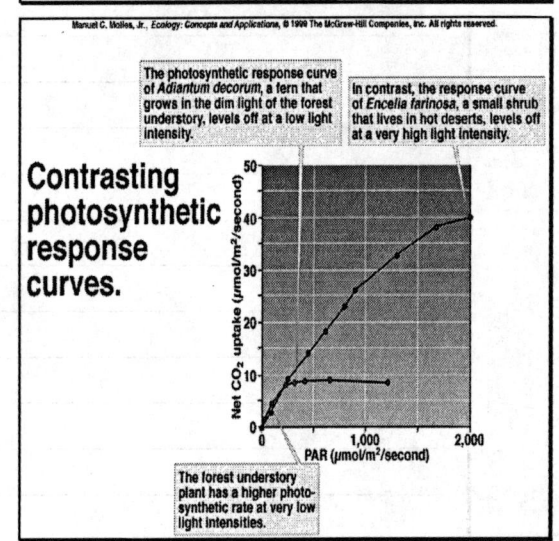

Contrasting photosynthetic response curves.

The photosynthetic response curve of *Adiantum decorum*, a fern that grows in the dim light of the forest understory, levels off at a low light intensity.

In contrast, the response curve of *Encelia farinosa*, a small shrub that lives in hot deserts, levels off at a very high light intensity.

The forest understory plant has a higher photosynthetic rate at very low light intensities.

52

Figure 6.21

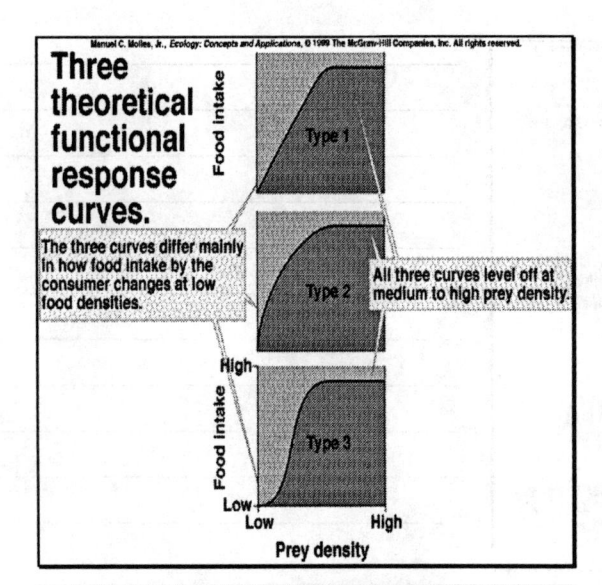

Three theoretical functional response curves.

The three curves differ mainly in how food intake by the consumer changes at low food densities.

All three curves level off at medium to high prey density.

Type 1
Type 2
Type 3

Figure 6.22

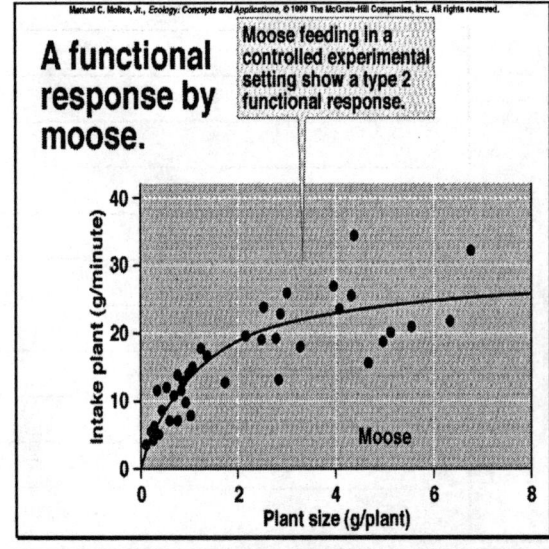

A functional response by moose.

Moose feeding in a controlled experimental setting show a type 2 functional response.

Figure 6.23

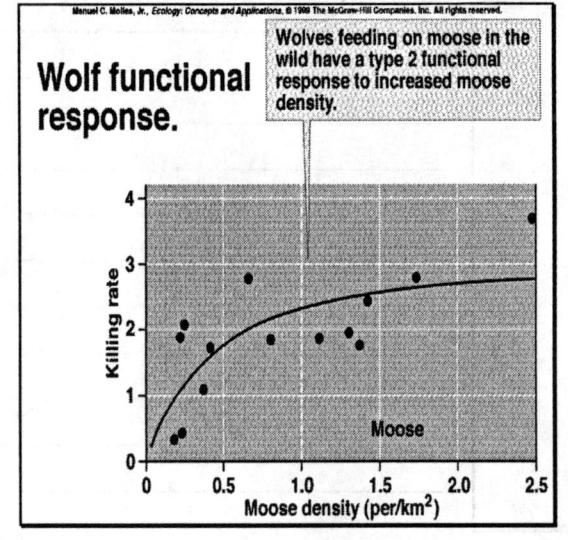

Wolf functional response.

Wolves feeding on moose in the wild have a type 2 functional response to increased moose density.

Figure 6.24

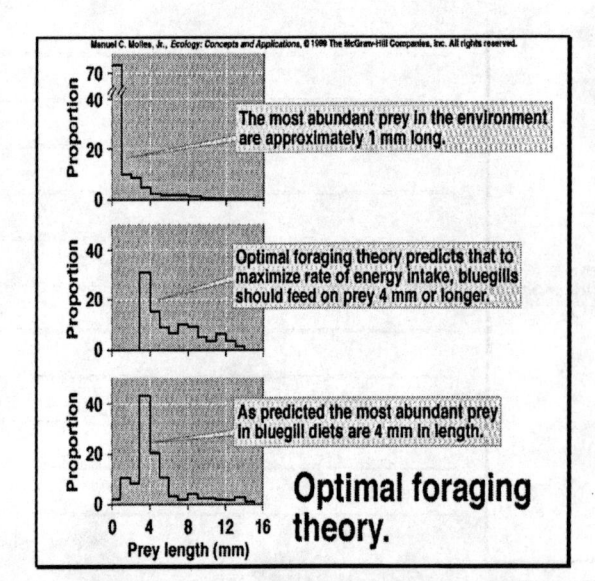

Proportion vs *Prey length (mm)*

- The most abundant prey in the environment are approximately 1 mm long.
- Optimal foraging theory predicts that to maximize rate of energy intake, bluegills should feed on prey 4 mm or longer.
- As predicted the most abundant prey in bluegill diets are 4 mm in length.

Optimal foraging theory.

Figure 6.25

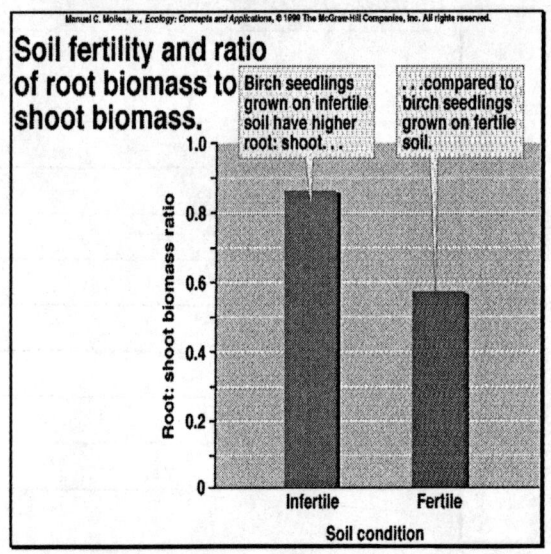

Soil fertility and ratio of root biomass to shoot biomass.

- Birch seedlings grown on infertile soil have higher root:shoot...
- ...compared to birch seedlings grown on fertile soil.

Figure 6.26

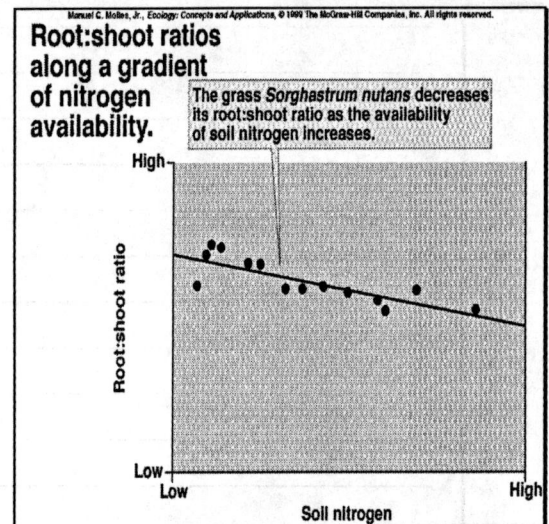

Root:shoot ratios along a gradient of nitrogen availability.

The grass *Sorghastrum nutans* decreases its root:shoot ratio as the availability of soil nitrogen increases.

54

Figure 6.27

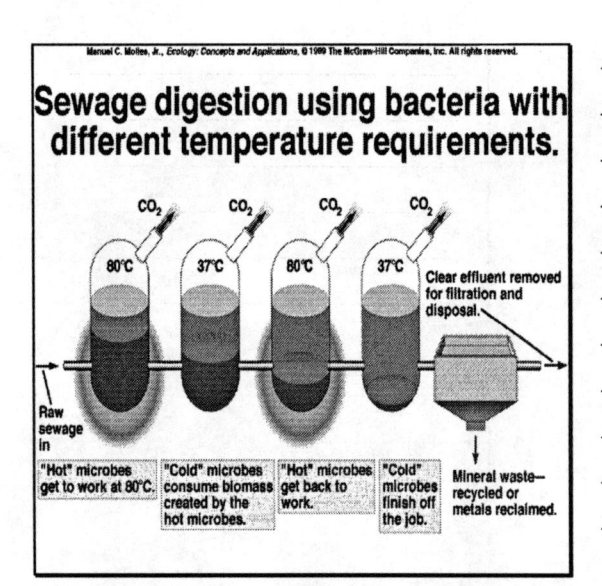

Sewage digestion using bacteria with different temperature requirements.

- Raw sewage in
- "Hot" microbes get to work at 80°C.
- "Cold" microbes consume biomass created by the hot microbes.
- "Hot" microbes get back to work.
- "Cold" microbes finish off the job.
- Clear effluent removed for filtration and disposal.
- Mineral waste—recycled or metals reclaimed.

Figure 6.28

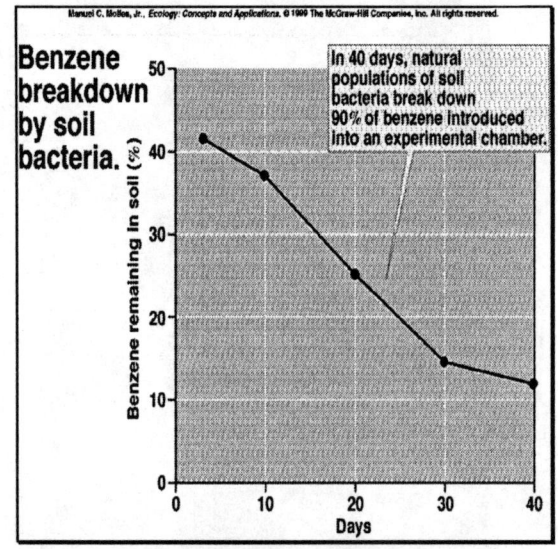

Benzene breakdown by soil bacteria.

In 40 days, natural populations of soil bacteria break down 90% of benzene introduced into an experimental chamber.

(Y-axis: Benzene remaining in soil (%); X-axis: Days)

Figure 6.29

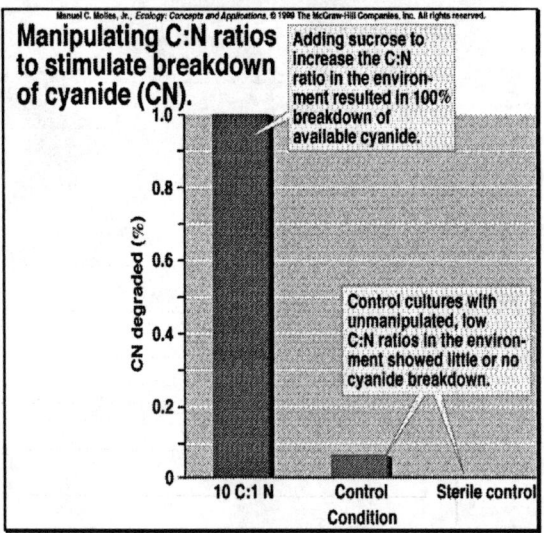

Manipulating C:N ratios to stimulate breakdown of cyanide (CN).

Adding sucrose to increase the C:N ratio in the environment resulted in 100% breakdown of available cyanide.

Control cultures with unmanipulated, low C:N ratios in the environment showed little or no cyanide breakdown.

(Y-axis: CN degraded (%); X-axis: Condition — 10 C:1 N, Control, Sterile control)

Chapter 7

Ecology

Concepts and Applications
First Edition

Manuel C. Molles, Jr.

(c) The McGraw-Hill Companies, Inc.

Figure 7.2

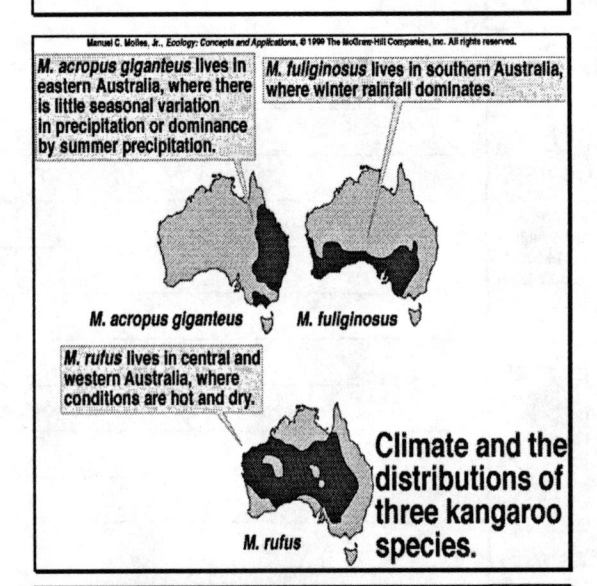

M. acropus giganteus lives in eastern Australia, where there is little seasonal variation in precipitation or dominance by summer precipitation.

M. fuliginosus lives in southern Australia, where winter rainfall dominates.

M. rufus lives in central and western Australia, where conditions are hot and dry.

Climate and the distributions of three kangaroo species.

Figure 7.3

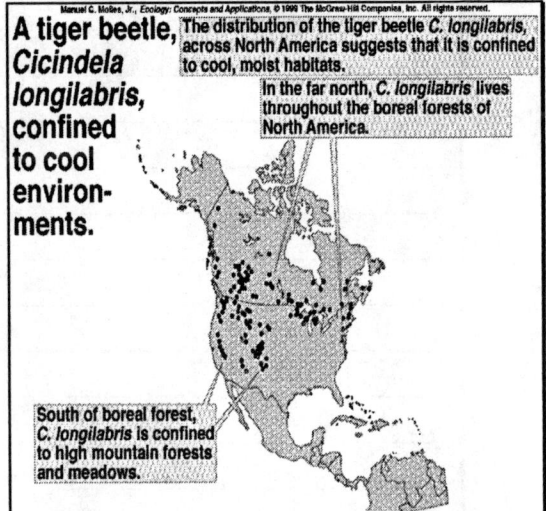

A tiger beetle, *Cicindela longilabris*, confined to cool environments.

The distribution of the tiger beetle *C. longilabris*, across North America suggests that it is confined to cool, moist habitats.

In the far north, *C. longilabris* lives throughout the boreal forests of North America.

South of boreal forest, *C. longilabris* is confined to high mountain forests and meadows.

Figure 7.4

Uniform temperature preference across an extensive geographic range.

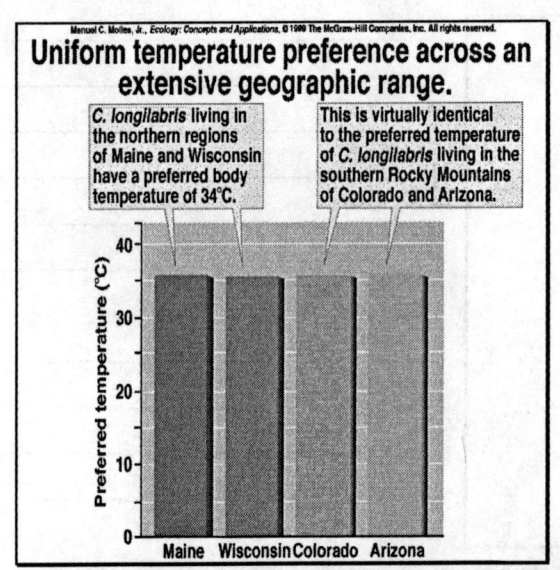

C. longilabris living in the northern regions of Maine and Wisconsin have a preferred body temperature of 34°C.

This is virtually identical to the preferred temperature of *C. longilabris* living in the southern Rocky Mountains of Colorado and Arizona.

Figure 7.5

The distributions of four *Encelia* species.

E. californica is confined to a narrow zone along the coast of California and Baja California, ... that is cool and moist in the north and cool and dry in the south.

E. actoni lives farther inland, where conditions are drier and slightly warmer than the areas inhabited by *E. californica*.

E. farinosa and *E. frutescens* live still farther inland in areas that are much hotter. The geographic distributions of these two species overlap a great deal.

E. californica   *E. actoni*   *E. farinosa*   *E. frutescens*

Figure 7.6

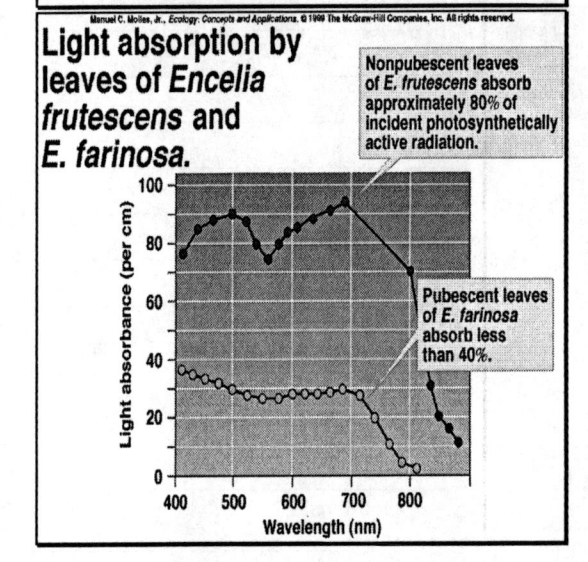

Light absorption by leaves of *Encelia frutescens* and *E. farinosa*.

Nonpubescent leaves of *E. frutescens* absorb approximately 80% of incident photosynthetically active radiation.

Pubescent leaves of *E. farinosa* absorb less than 40%.

58

Figure 7.7

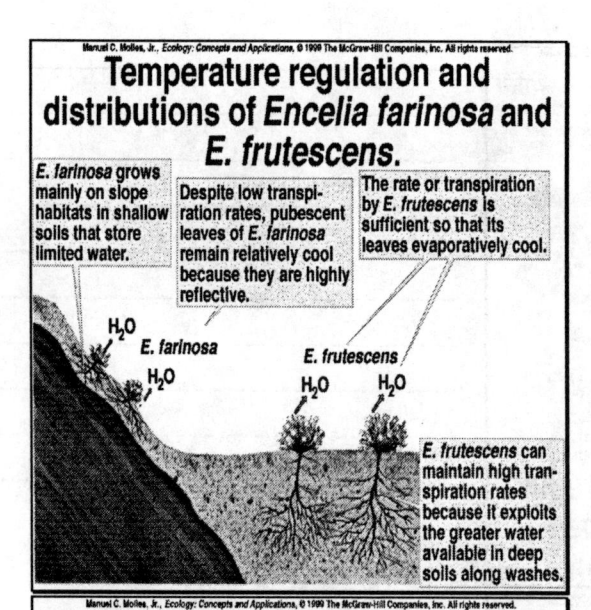

Figure 7.8

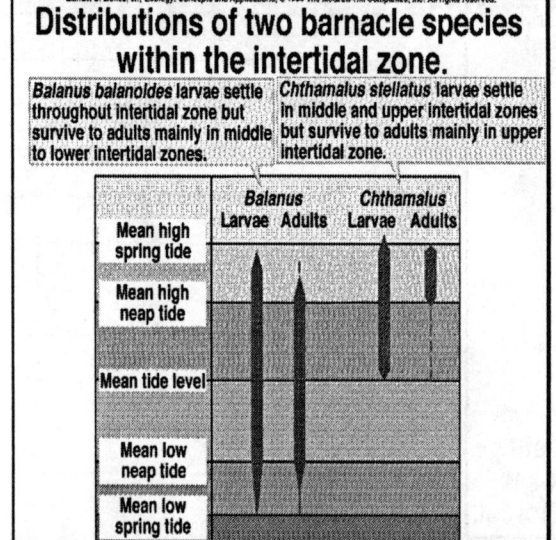

Figure 7.9

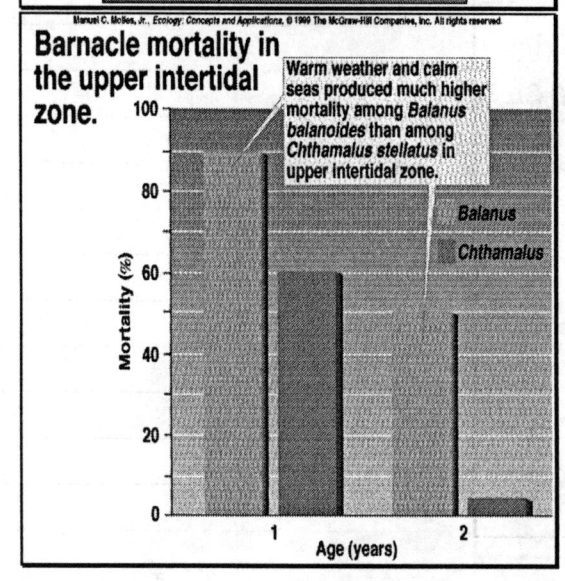

Figure 7.10

Random, regular, and clumped distributions.

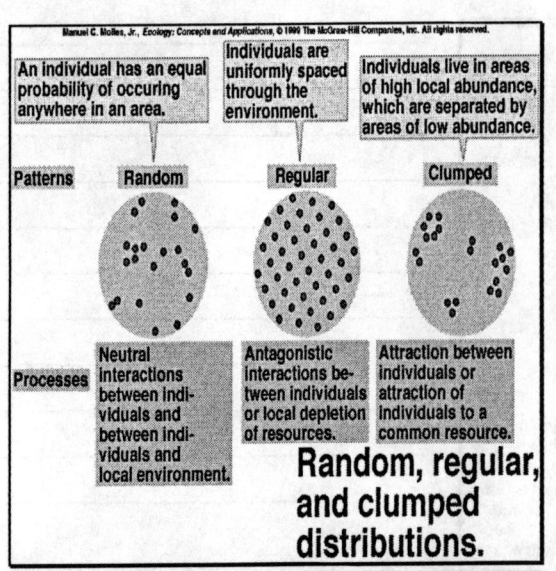

Patterns: Random — An individual has an equal probability of occuring anywhere in an area. Regular — Individuals are uniformly spaced through the environment. Clumped — Individuals live in areas of high local abundance, which are separated by areas of low abundance.

Processes: Neutral interactions between individuals and between individuals and local environment. Antagonistic interactions between individuals or local depletion of resources. Attraction between individuals or attraction of individuals to a common resource.

Figure 7.11

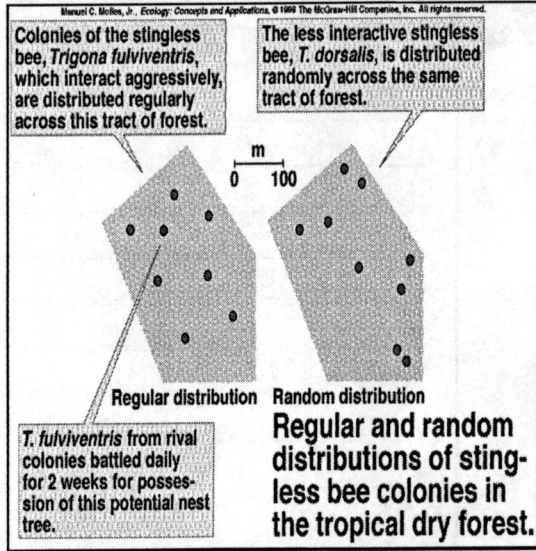

Regular and random distributions of stingless bee colonies in the tropical dry forest.

Colonies of the stingless bee, *Trigona fulviventris*, which interact aggressively, are distributed regularly across this tract of forest.

The less interactive stingless bee, *T. dorsalis*, is distributed randomly across the same tract of forest.

T. fulviventris from rival colonies battled daily for 2 weeks for possession of this potential nest tree.

Figure 7.13

Hypothetical change in shrub distributions with increasing shrub size.

Small shrubs establish in high densities and produce a clumped distribution. — Young, small shrubs — Clumped

Mortality as the shrubs grow reduces clumping and produces a random distribution among medium shrubs. — Medium shrubs — Random

Competition enforces a regular distribution among large shrubs. — Large shrubs — Regular

Figure 7.14

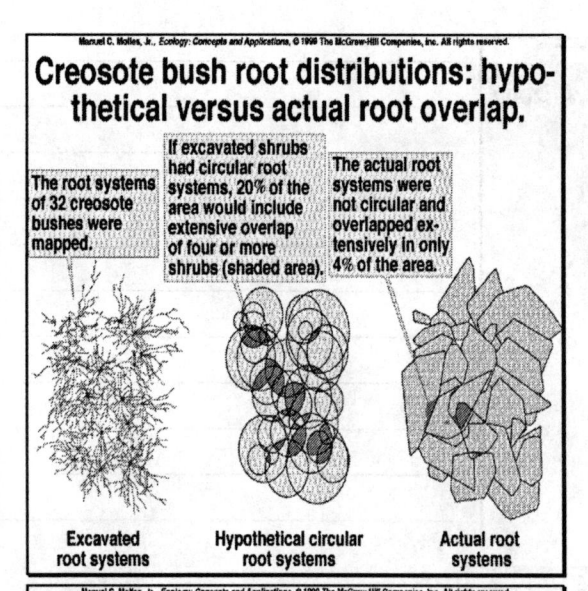

Creosote bush root distributions: hypothetical versus actual root overlap.

The root systems of 32 creosote bushes were mapped.

If excavated shrubs had circular root systems, 20% of the area would include extensive overlap of four or more shrubs (shaded area).

The actual root systems were not circular and overlapped extensively in only 4% of the area.

Excavated root systems | Hypothetical circular root systems | Actual root systems

Figure 7.15

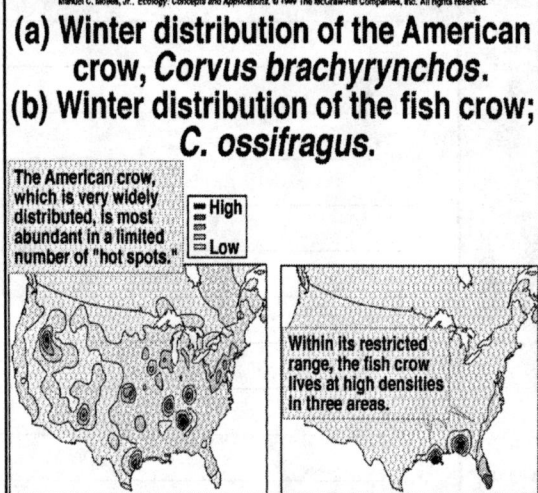

(a) Winter distribution of the American crow, *Corvus brachyrynchos*.
(b) Winter distribution of the fish crow; *C. ossifragus*.

The American crow, which is very widely distributed, is most abundant in a limited number of "hot spots."

High
Low

Within its restricted range, the fish crow lives at high densities in three areas.

(a) (b)

Figure 7.16

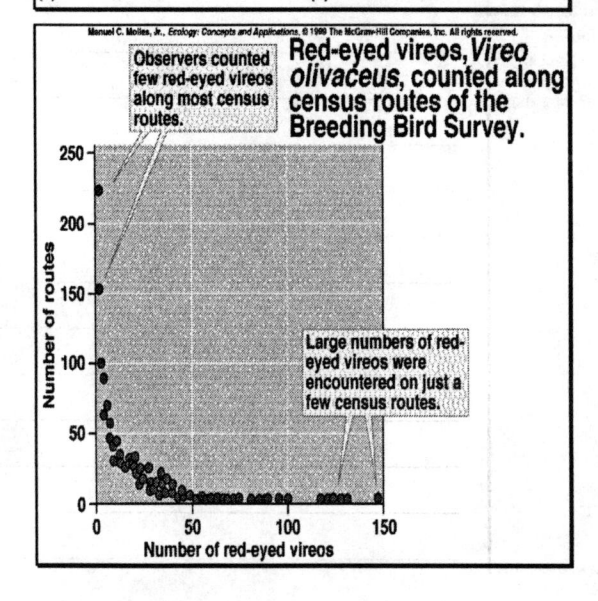

Red-eyed vireos, *Vireo olivaceus*, counted along census routes of the Breeding Bird Survey.

Observers counted few red-eyed vireos along most census routes.

Large numbers of red-eyed vireos were encountered on just a few census routes.

Number of routes vs. Number of red-eyed vireos

Figure 7.17

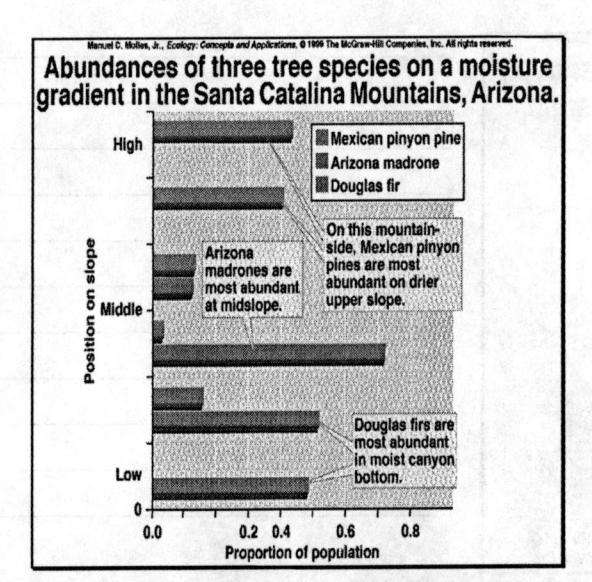

Abundances of three tree species on a moisture gradient in the Santa Catalina Mountains, Arizona.

- Mexican pinyon pine
- Arizona madrone
- Douglas fir

On this mountainside, Mexican pinyon pines are most abundant on drier upper slope.

Arizona madrones are most abundant at midslope.

Douglas firs are most abundant in moist canyon bottom.

Figure 7.18

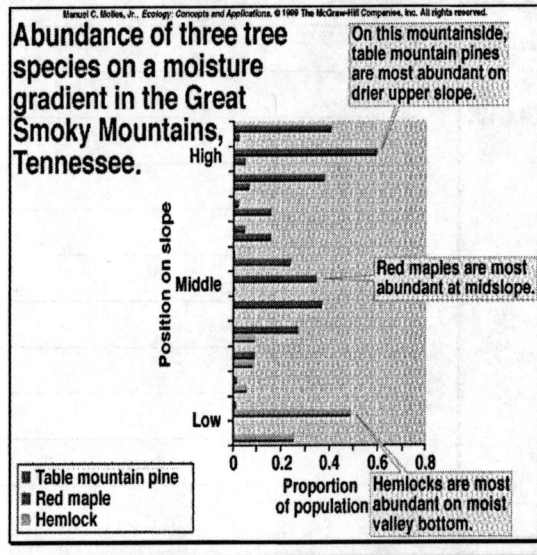

Abundance of three tree species on a moisture gradient in the Great Smoky Mountains, Tennessee.

- Table mountain pine
- Red maple
- Hemlock

On this mountainside, table mountain pines are most abundant on drier upper slope.

Red maples are most abundant at midslope.

Hemlocks are most abundant on moist valley bottom.

Figure 7.19

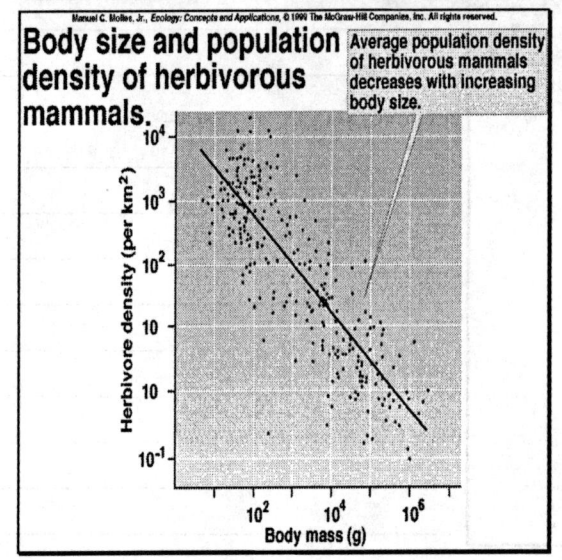

Body size and population density of herbivorous mammals.

Average population density of herbivorous mammals decreases with increasing body size.

Figure 7.20

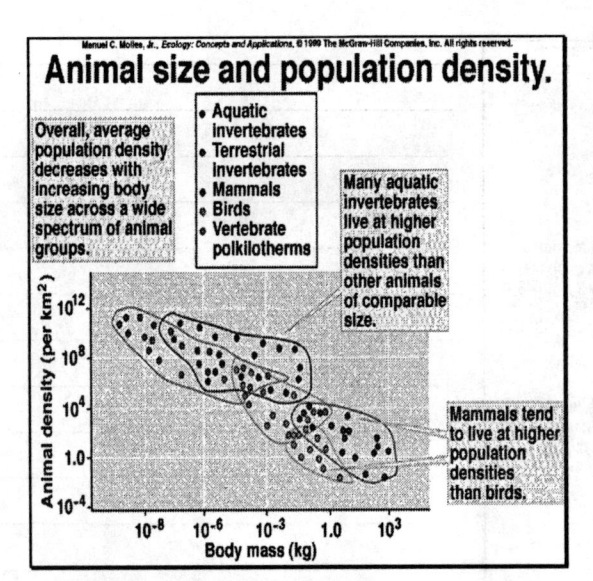

Figure 7.21

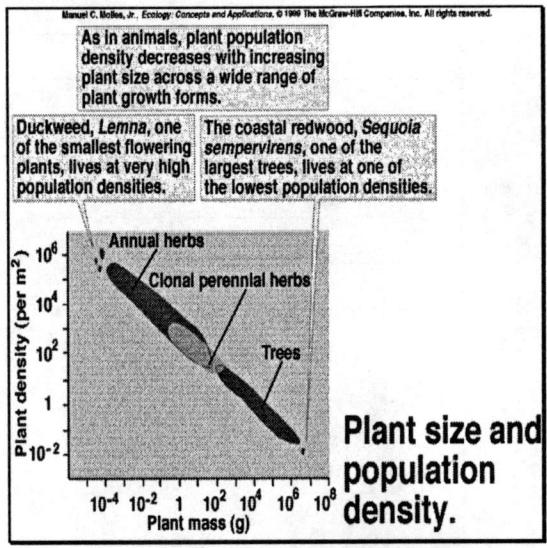

Figure 7.22

Figure 7.27

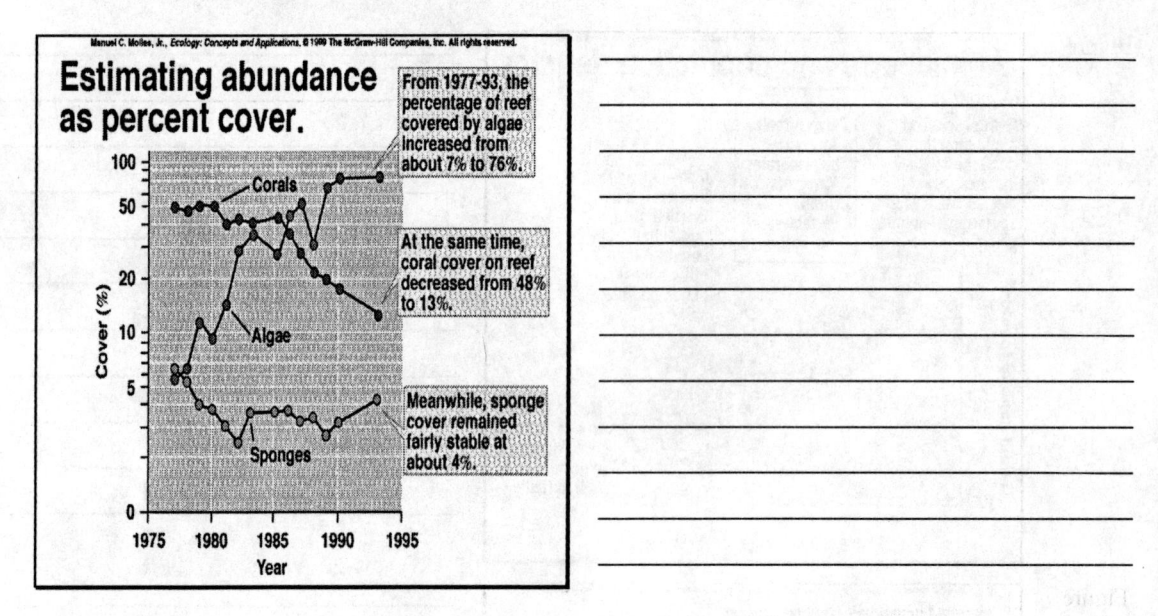

Chapter 8

Ecology

Concepts and Applications
First Edition

Manuel C. Molles, Jr.

(c) The McGraw-Hill Companies, Inc.

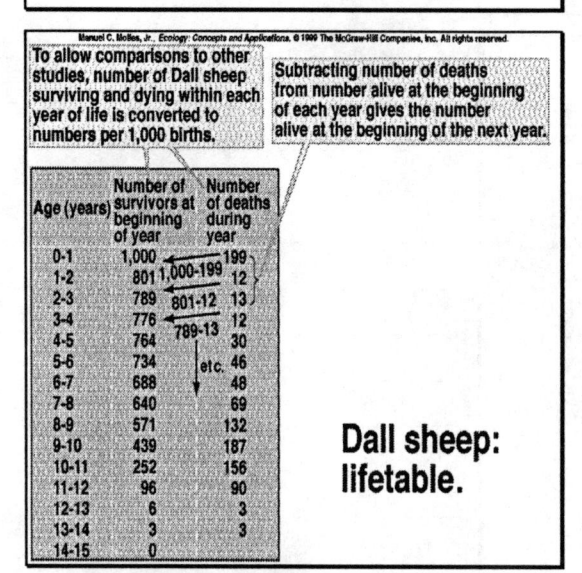

Figure 8.2

Dall sheep: lifetable.

To allow comparisons to other studies, number of Dall sheep surviving and dying within each year of life is converted to numbers per 1,000 births.

Subtracting number of deaths from number alive at the beginning of each year gives the number alive at the beginning of the next year.

Age (years)	Number of survivors at beginning of year	Number of deaths during year
0-1	1,000	199
1-2	801	12
2-3	789	13
3-4	776	12
4-5	764	30
5-6	734	46
6-7	688	48
7-8	640	69
8-9	571	132
9-10	439	187
10-11	252	156
11-12	96	90
12-13	6	3
13-14	3	3
14-15	0	

Figure 8.2

Dall sheep: survivorship curve.

Plotting age on the x-axis and number of survivors on the y-axis creates a survivorship curve.

Dall sheep surviving their first year of life have a high probability of surviving to about age 9.

Sheep 10 years old and older are easier prey for wolves and die at a high rate.

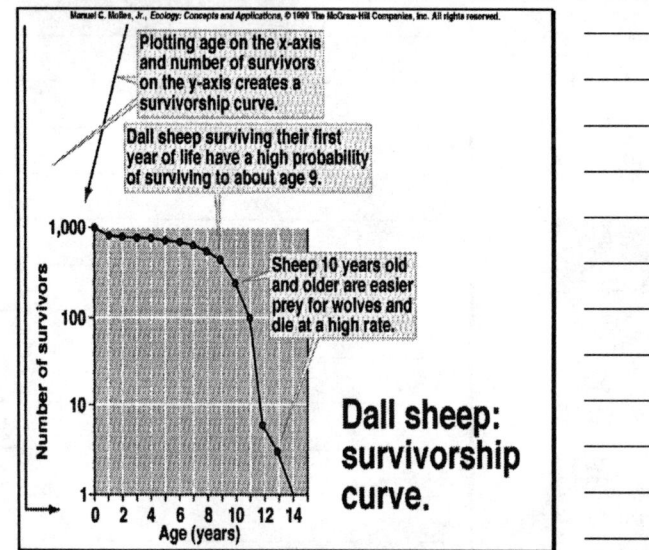

Figure 8.3

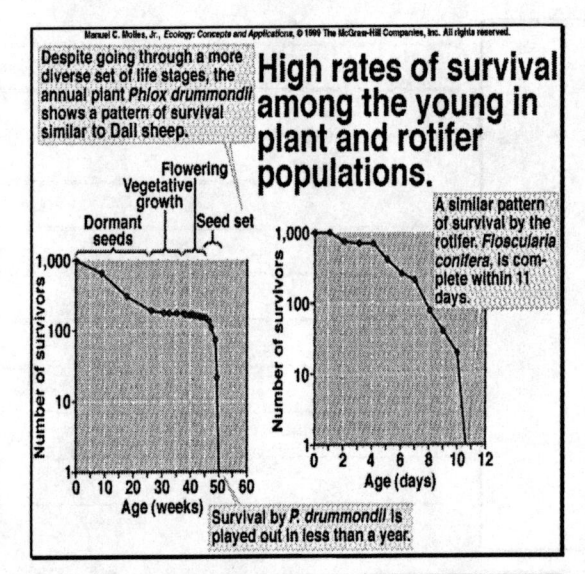

High rates of survival among the young in plant and rotifer populations.

Despite going through a more diverse set of life stages, the annual plant *Phlox drummondii* shows a pattern of survival similar to Dall sheep.

A similar pattern of survival by the rotifer, *Floscularia conifera*, is complete within 11 days.

Survival by *P. drummondii* is played out in less than a year.

Figure 8.4

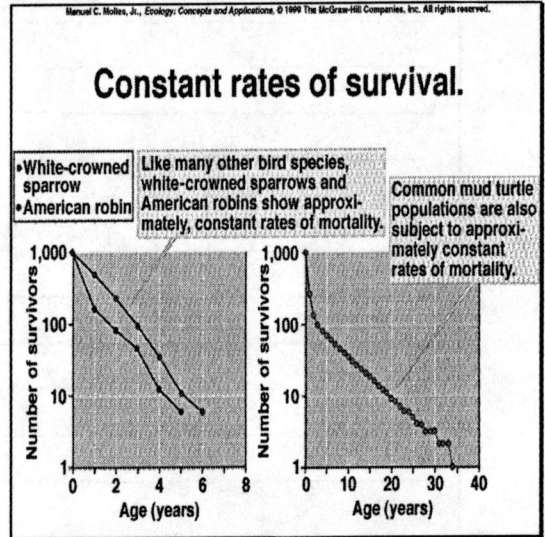

Constant rates of survival.

- White-crowned sparrow
- American robin

Like many other bird species, white-crowned sparrows and American robins show approximately, constant rates of mortality.

Common mud turtle populations are also subject to approximately constant rates of mortality.

Figure 8.5

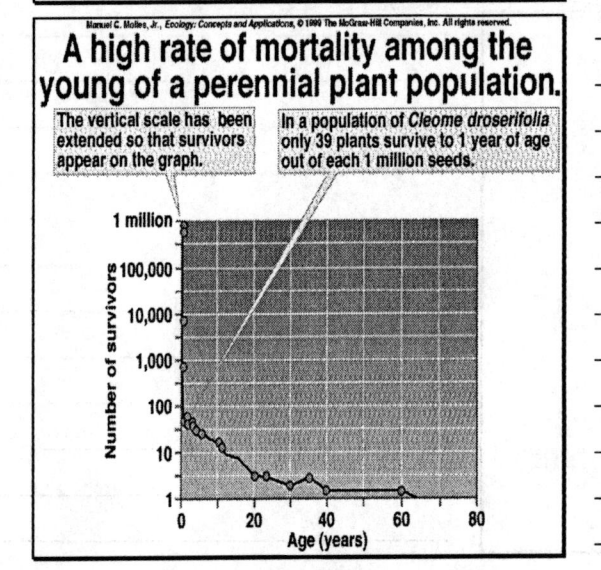

A high rate of mortality among the young of a perennial plant population.

The vertical scale has been extended so that survivors appear on the graph.

In a population of *Cleome droserifolia* only 39 plants survive to 1 year of age out of each 1 million seeds.

Figure 8.6

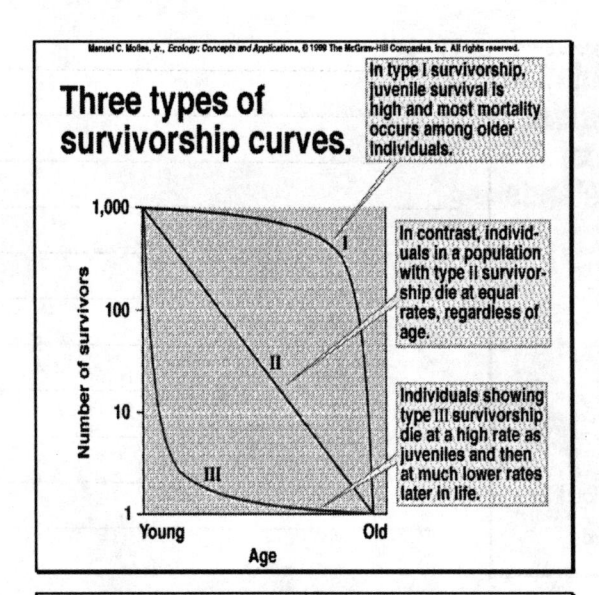

Three types of survivorship curves.

In type I survivorship, juvenile survival is high and most mortality occurs among older individuals.

In contrast, individuals in a population with type II survivorship die at equal rates, regardless of age.

Individuals showing type III survivorship die at a high rate as juveniles and then at much lower rates later in life.

Figure 8.7

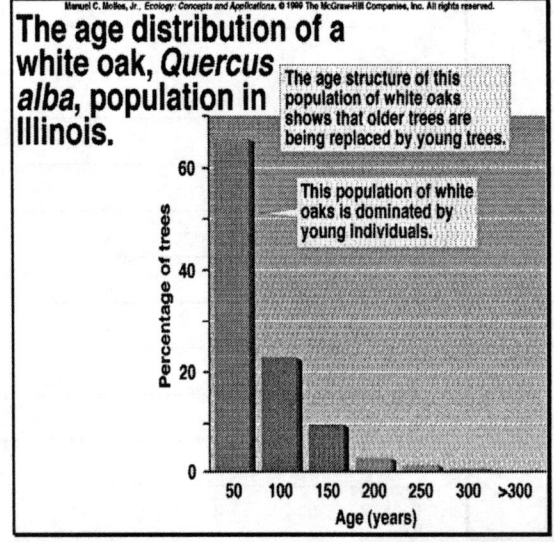

The age distribution of a white oak, *Quercus alba*, population in Illinois.

The age structure of this population of white oaks shows that older trees are being replaced by young trees.

This population of white oaks is dominated by young individuals.

Figure 8.8

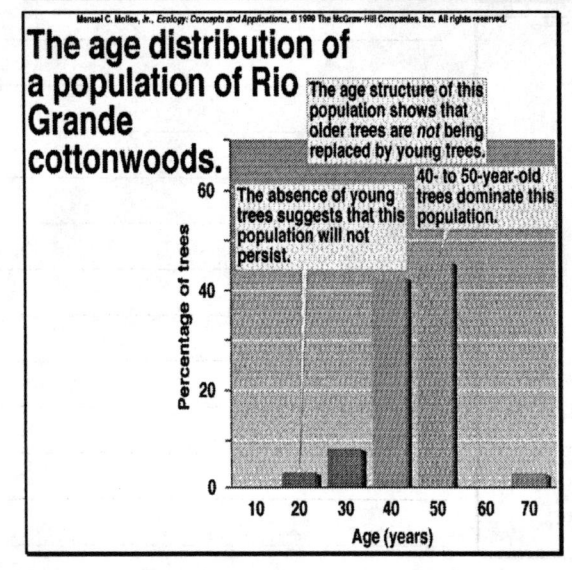

The age distribution of a population of Rio Grande cottonwoods.

The age structure of this population shows that older trees are *not* being replaced by young trees.

The absence of young trees suggests that this population will not persist.

40- to 50-year-old trees dominate this population.

Figure 8.9

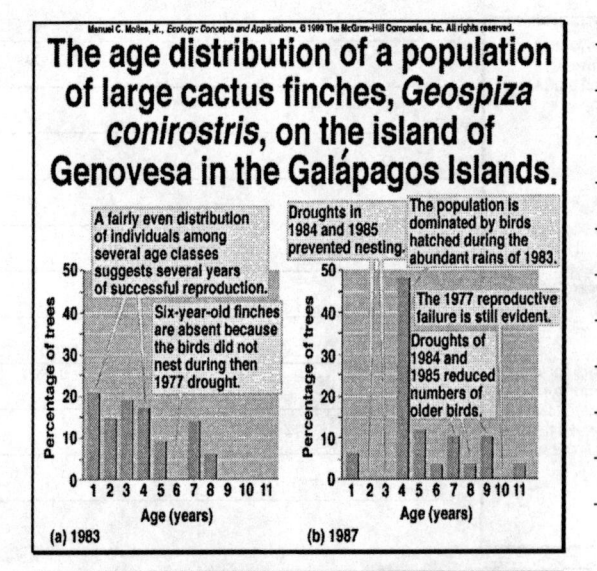

Figure 8.10

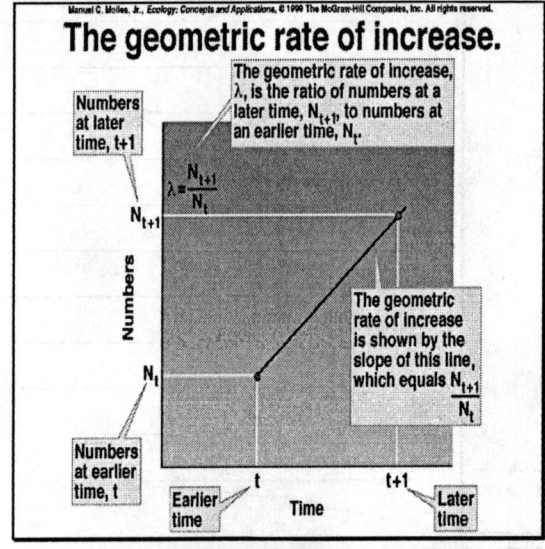

Figure 8.11

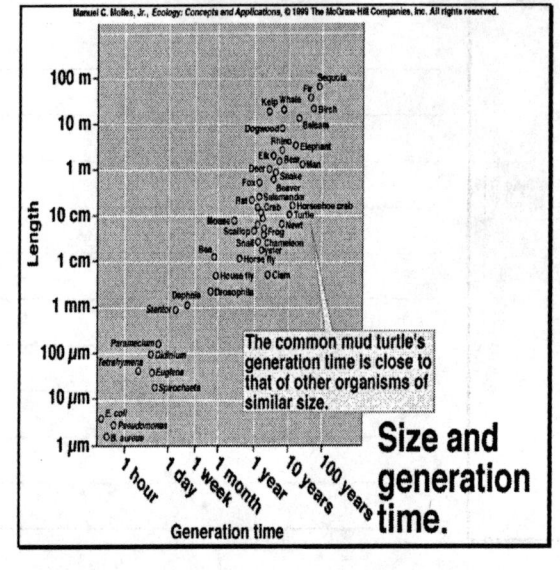

Figure 8.12

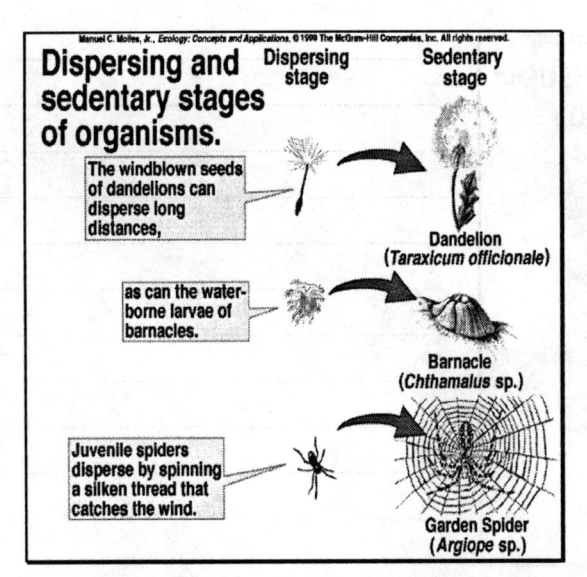

Dispersing and sedentary stages of organisms.

Dispersing stage — Sedentary stage

The windblown seeds of dandelions can disperse long distances,
Dandelion (*Taraxicum officionale*)

as can the water-borne larvae of barnacles.
Barnacle (*Chthamalus* sp.)

Juvenile spiders disperse by spinning a silken thread that catches the wind.
Garden Spider (*Argiope* sp.)

Figure 8.13

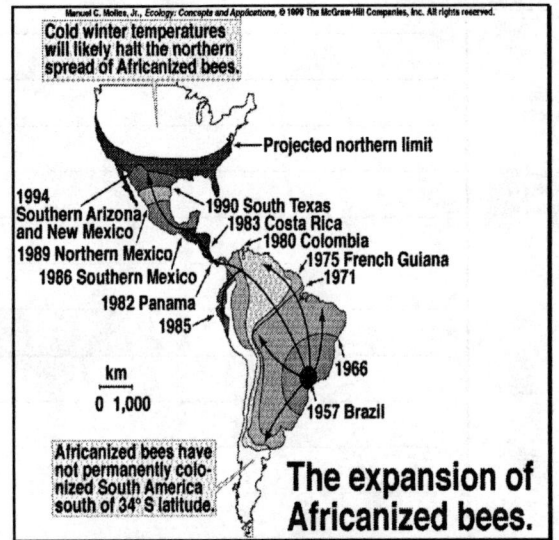

Cold winter temperatures will likely halt the northern spread of Africanized bees.

Projected northern limit
1994 Southern Arizona and New Mexico
1990 South Texas
1989 Northern Mexico
1983 Costa Rica
1986 Southern Mexico
1980 Colombia
1982 Panama
1975 French Guiana
1985
1971
1966
1957 Brazil

km 0 1,000

Africanized bees have not permanently colonized South America south of 34° S latitude.

The expansion of Africanized bees.

Figure 8.14

The expansion of collared doves, *Streptopelia decaocto*.

In less than 60 years collared doves expanded their range to the farthest corners of Europe.

1990, 1980, 1970, 1960, 1990, 1950, 1940, 1930, 1900

Collared doves began to spread out of Turkey into Europe early in the twentieth century.

69

Figure 8.15

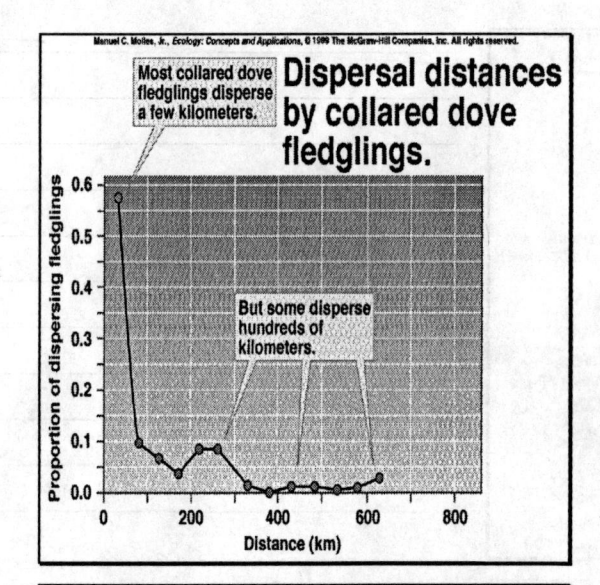

Dispersal distances by collared dove fledglings.

Most collared dove fledglings disperse a few kilometers.

But some disperse hundreds of kilometers.

Figure 8.16

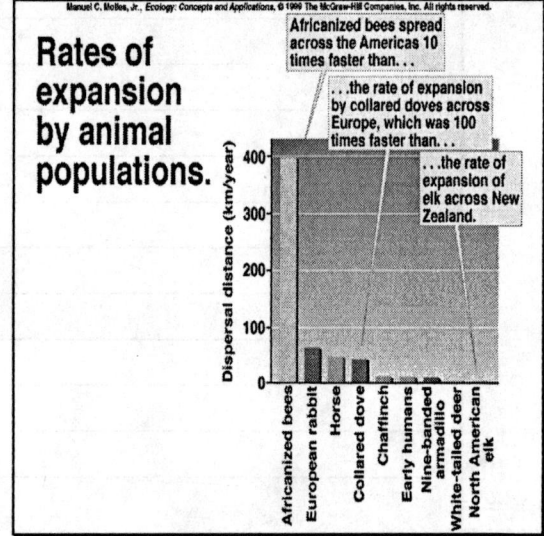

Rates of expansion by animal populations.

Africanized bees spread across the Americas 10 times faster than...

...the rate of expansion by collared doves across Europe, which was 100 times faster than...

...the rate of expansion of elk across New Zealand.

Figure 8.17

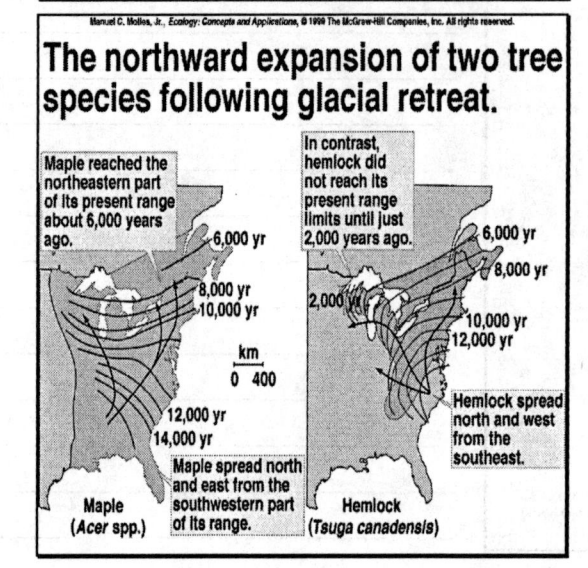

The northward expansion of two tree species following glacial retreat.

Maple reached the northeastern part of its present range about 6,000 years ago.

In contrast, hemlock did not reach its present range limits until just 2,000 years ago.

Maple spread north and east from the southwestern part of its range.

Hemlock spread north and west from the southeast.

Maple (*Acer* spp.)

Hemlock (*Tsuga canadensis*)

Figure 8.18

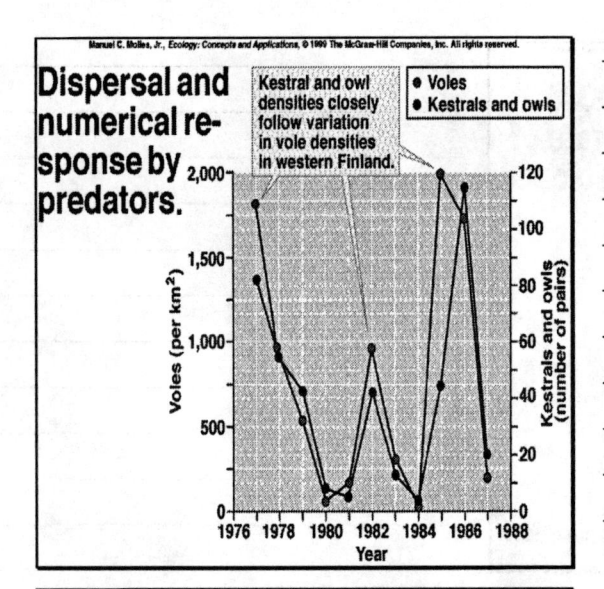

Dispersal and numerical response by predators.

Kestral and owl densities closely follow variation in vole densities in western Finland.

- Voles
- Kestrals and owls

Figure 8.19

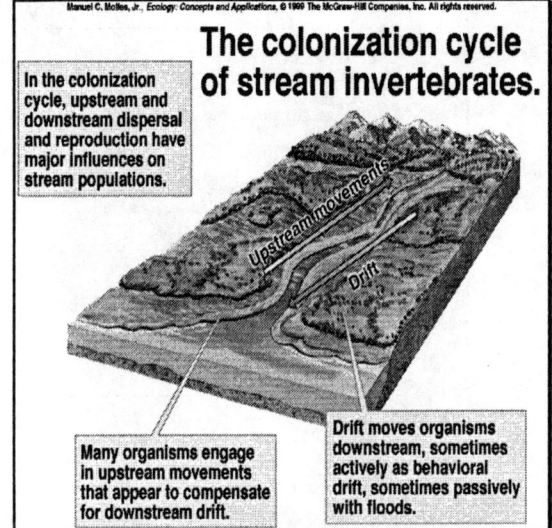

The colonization cycle of stream invertebrates.

In the colonization cycle, upstream and downstream dispersal and reproduction have major influences on stream populations.

Upstream movements

Drift

Many organisms engage in upstream movements that appear to compensate for downstream drift.

Drift moves organisms downstream, sometimes actively as behavioral drift, sometimes passively with floods.

Figure 8.21

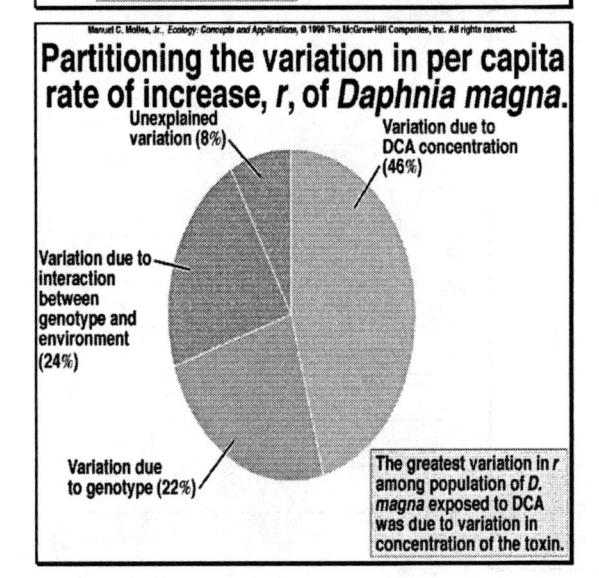

Partitioning the variation in per capita rate of increase, r, of *Daphnia magna*.

- Unexplained variation (8%)
- Variation due to DCA concentration (46%)
- Variation due to interaction between genotype and environment (24%)
- Variation due to genotype (22%)

The greatest variation in r among population of *D. magna* exposed to DCA was due to variation in concentration of the toxin.

Figure 8.22

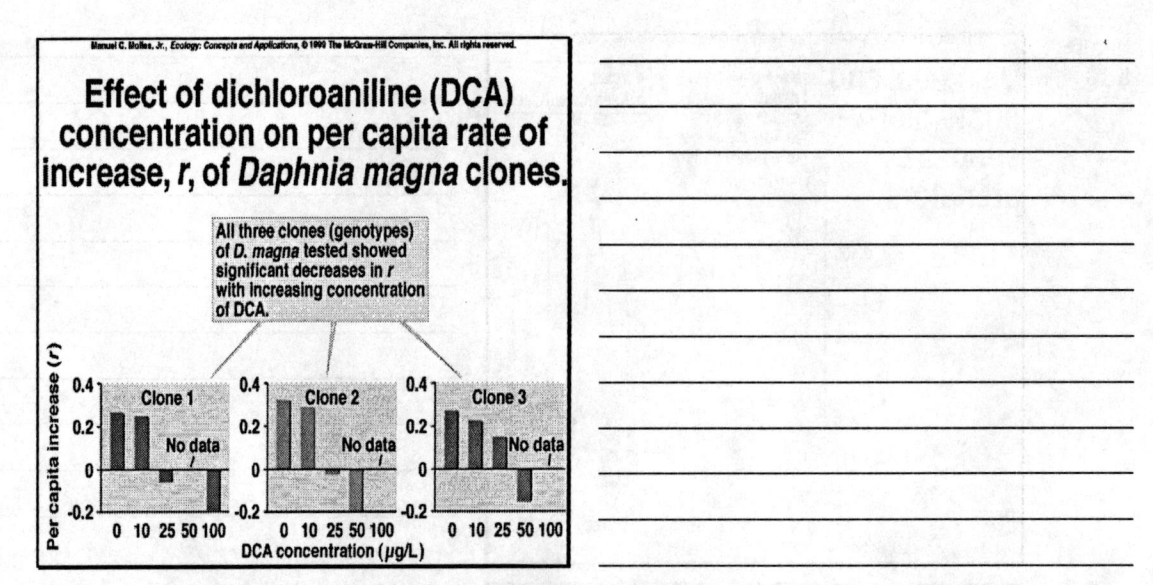

Chapter 9

Ecology
Concepts and Applications
First Edition

Manuel C. Molles, Jr.

(c) The McGraw-Hill Companies, Inc.

Figure 9.2

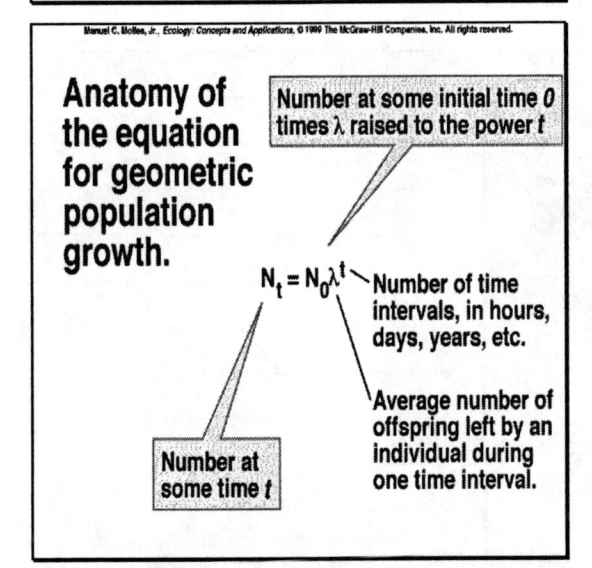

Anatomy of the equation for geometric population growth.

$N_t = N_0 \lambda^t$

- Number at some initial time 0 times λ raised to the power t
- Number at some time t
- Number of time intervals, in hours, days, years, etc.
- Average number of offspring left by an individual during one time interval.

Figure 9.3

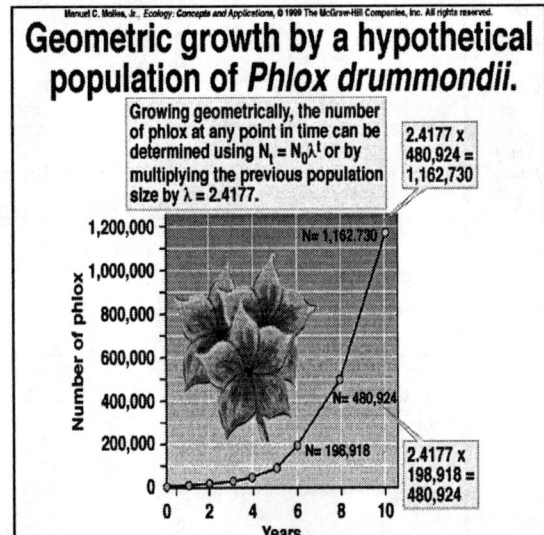

Geometric growth by a hypothetical population of *Phlox drummondii*.

Growing geometrically, the number of phlox at any point in time can be determined using $N_t = N_0 \lambda^t$ or by multiplying the previous population size by $\lambda = 2.4177$.

2.4177 × 480,924 = 1,162,730

N = 1,162,730
N = 480,924
N = 198,918

2.4177 × 198,918 = 480,924

Figure 9.4

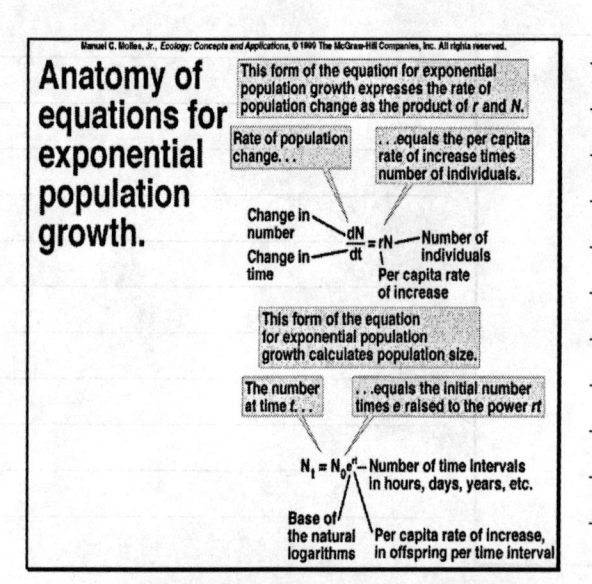

Anatomy of equations for exponential population growth.

This form of the equation for exponential population growth expresses the rate of population change as the product of r and N.

Rate of population change... ...equals the per capita rate of increase times number of individuals.

$$\frac{dN}{dt} = rN$$

Change in number — dN
Change in time — dt
Number of individuals — N
Per capita rate of increase — r

This form of the equation for exponential population growth calculates population size.

The number at time t... ...equals the initial number times e raised to the power rt.

$$N_t = N_0 e^{rt}$$

- N_t — Number at time t
- N_0 — Initial number
- e — Base of the natural logarithms
- r — Per capita rate of increase, in offspring per time interval
- t — Number of time intervals in hours, days, years, etc.

Figure 9.5

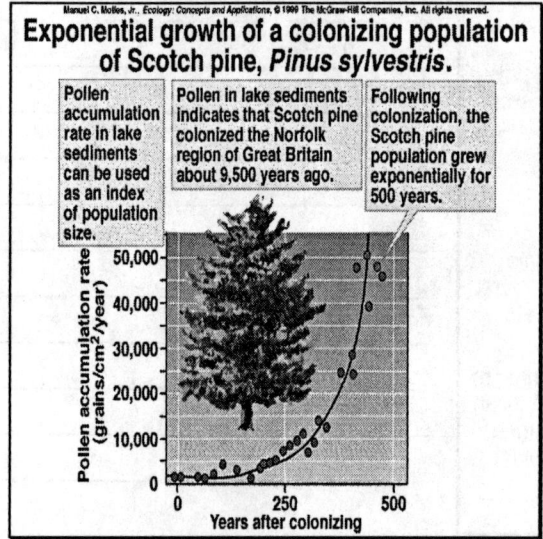

Exponential growth of a colonizing population of Scotch pine, *Pinus sylvestris*.

Pollen accumulation rate in lake sediments can be used as an index of population size.

Pollen in lake sediments indicates that Scotch pine colonized the Norfolk region of Great Britain about 9,500 years ago.

Following colonization, the Scotch pine population grew exponentially for 500 years.

(Graph: Pollen accumulation rate (grains/cm²/year) vs. Years after colonizing, 0 to 500; values rising from 0 to ~50,000.)

Figure 9.6

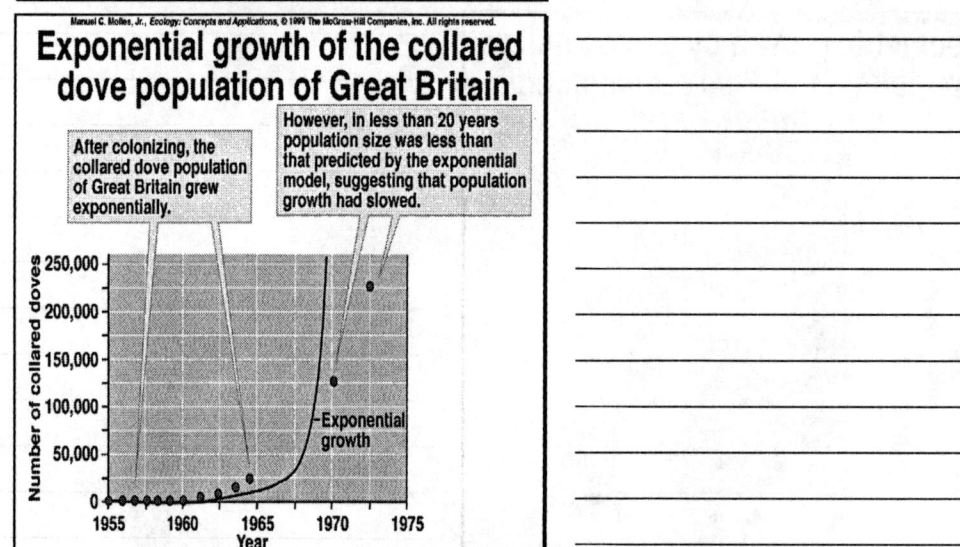

Exponential growth of the collared dove population of Great Britain.

After colonizing, the collared dove population of Great Britain grew exponentially.

However, in less than 20 years population size was less than that predicted by the exponential model, suggesting that population growth had slowed.

(Graph: Number of collared doves vs. Year, 1955 to 1975; exponential growth curve.)

Figure 9.7

Sigmoidal population growth.

Carrying Capacity: Theoretical maximum population

Population grows rapidly.

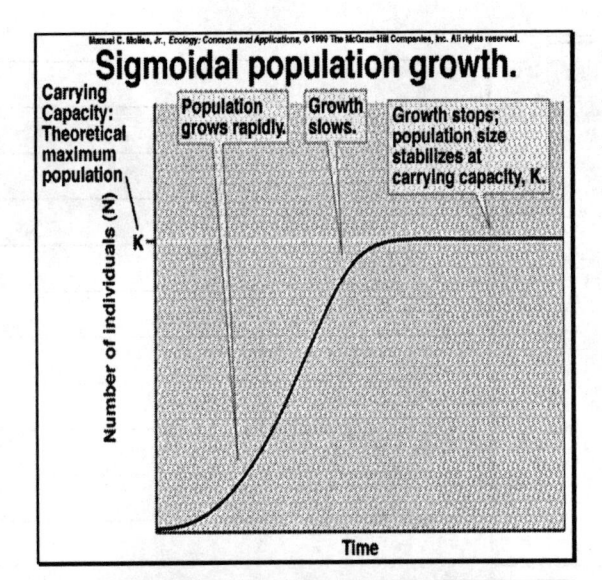

Growth slows.

Growth stops; population size stabilizes at carrying capacity, K.

Number of individuals (N)

Time

Figure 9.8

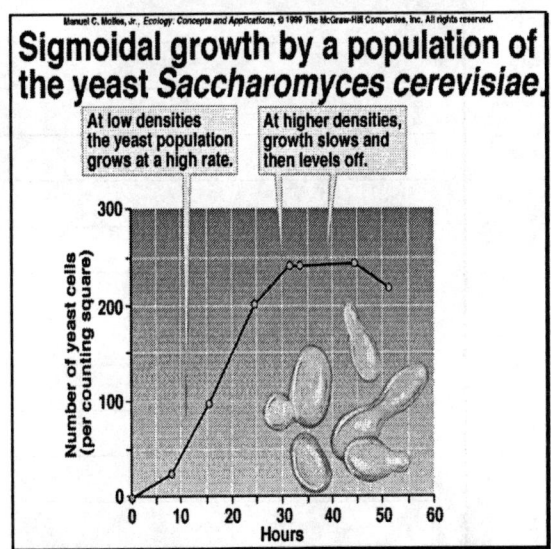

Sigmoidal growth by a population of the yeast *Saccharomyces cerevisiae*.

At low densities the yeast population grows at a high rate.

At higher densities, growth slows and then levels off.

Number of yeast cells (per counting square)

Hours

Figure 9.9

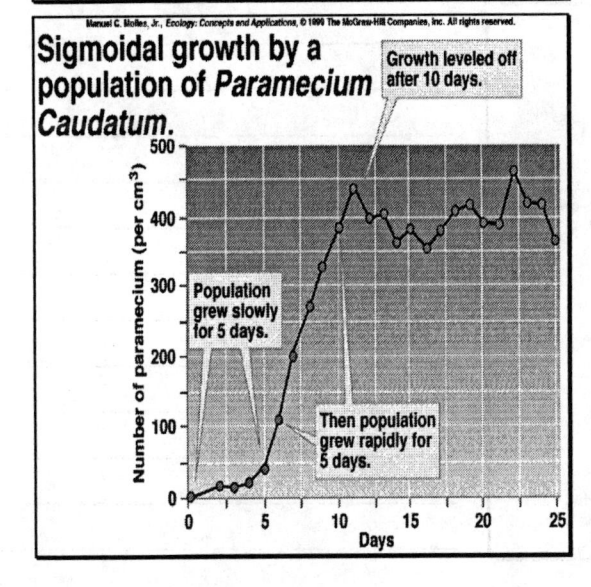

Sigmoidal growth by a population of *Paramecium Caudatum*.

Growth leveled off after 10 days.

Population grew slowly for 5 days.

Then population grew rapidly for 5 days.

Number of paramecium (per cm^3)

Days

75

Figure 9.10

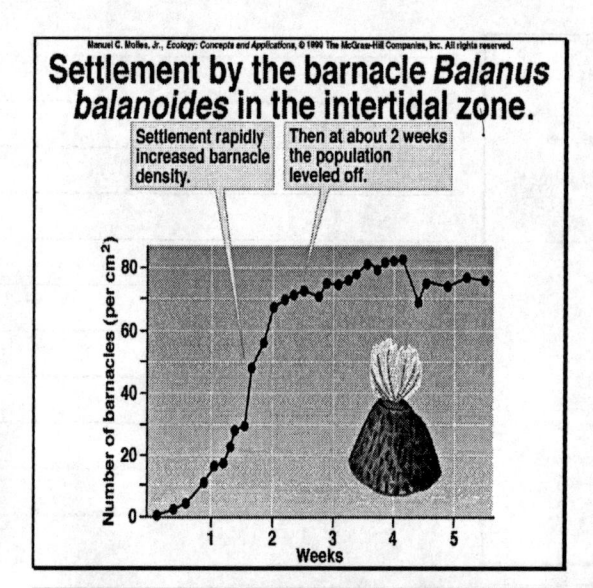

Settlement by the barnacle *Balanus balanoides* in the intertidal zone.

Settlement rapidly increased barnacle density. Then at about 2 weeks the population leveled off.

Figure 9.11

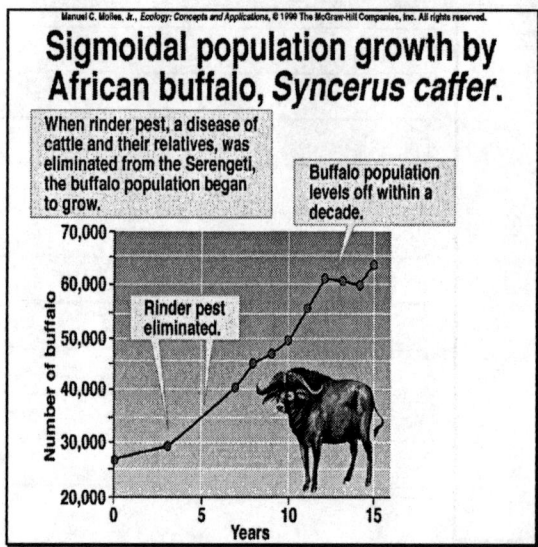

Sigmoidal population growth by African buffalo, *Syncerus caffer*.

When rinder pest, a disease of cattle and their relatives, was eliminated from the Serengeti, the buffalo population began to grow.

Buffalo population levels off within a decade.

Rinder pest eliminated.

Figure 9.12

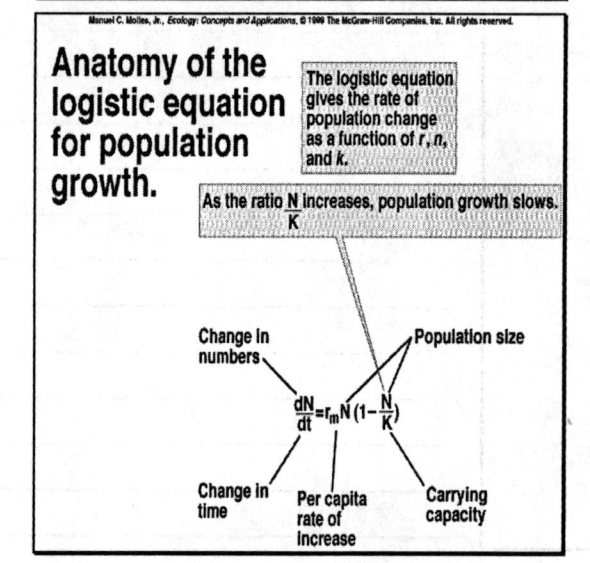

Anatomy of the logistic equation for population growth.

The logistic equation gives the rate of population change as a function of r, n, and k.

As the ratio $\frac{N}{K}$ increases, population growth slows.

$$\frac{dN}{dt} = r_m N \left(1 - \frac{N}{K}\right)$$

Change in numbers — Population size
Change in time — Per capita rate of increase — Carrying capacity

Figure 9.13

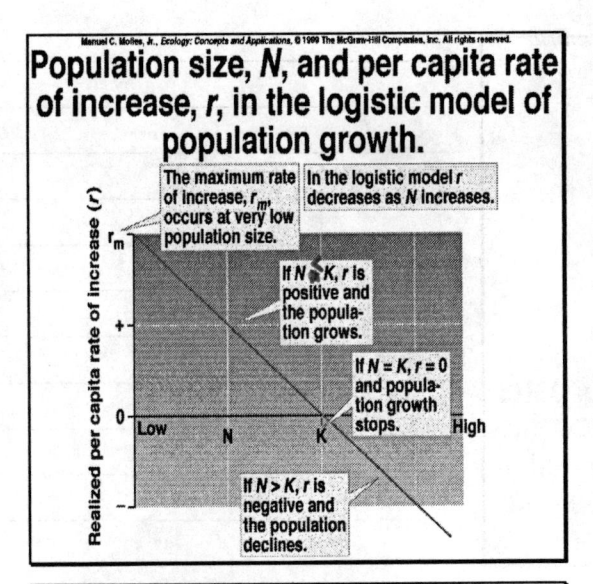

Population size, N, and per capita rate of increase, r, in the logistic model of population growth.

- The maximum rate of increase, r_m, occurs at very low population size.
- In the logistic model r decreases as N increases.
- If $N < K$, r is positive and the population grows.
- If $N = K$, $r = 0$ and population growth stops.
- If $N > K$, r is negative and the population declines.

Figure 9.14

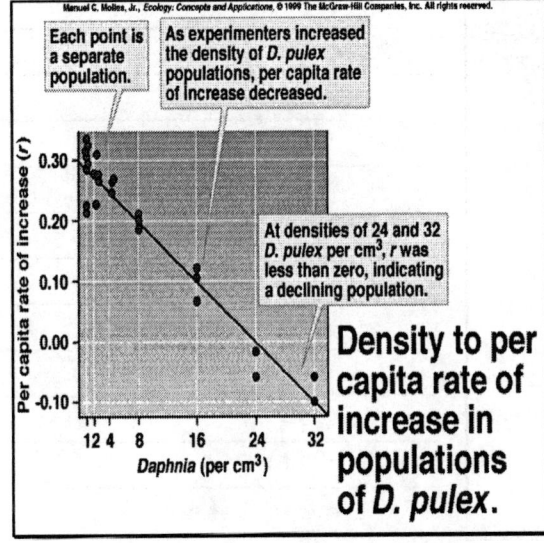

Density to per capita rate of increase in populations of *D. pulex*.

- Each point is a separate population.
- As experimenters increased the density of *D. pulex* populations, per capita rate of increase decreased.
- At densities of 24 and 32 *D. pulex* per cm³, r was less than zero, indicating a declining population.

Figure 9.16

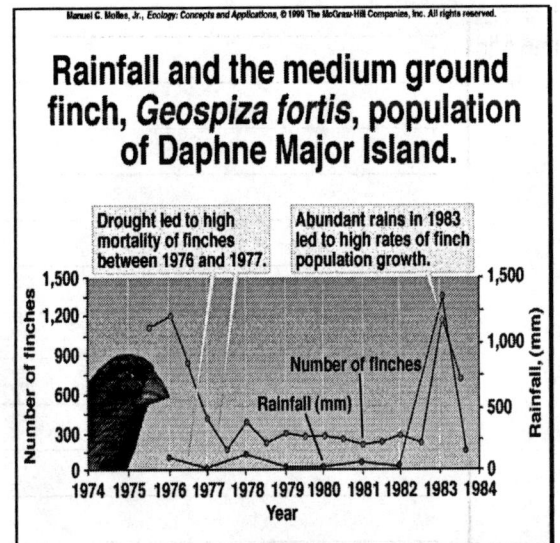

Rainfall and the medium ground finch, *Geospiza fortis*, population of Daphne Major Island.

- Drought led to high mortality of finches between 1976 and 1977.
- Abundant rains in 1983 led to high rates of finch population growth.

77

Figure 9.17

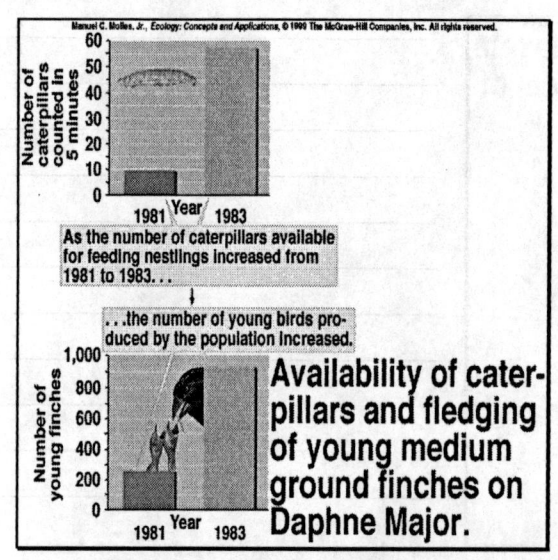

Availability of caterpillars and fledging of young medium ground finches on Daphne Major.

Figure 9.18

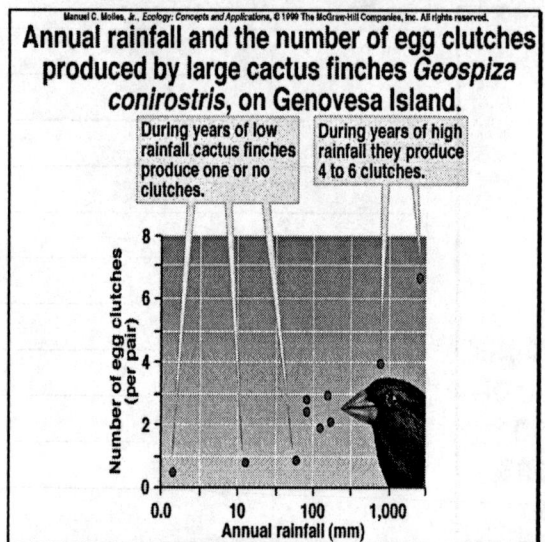

Annual rainfall and the number of egg clutches produced by large cactus finches *Geospiza conirostris*, on Genovesa Island.

Figure 9.20

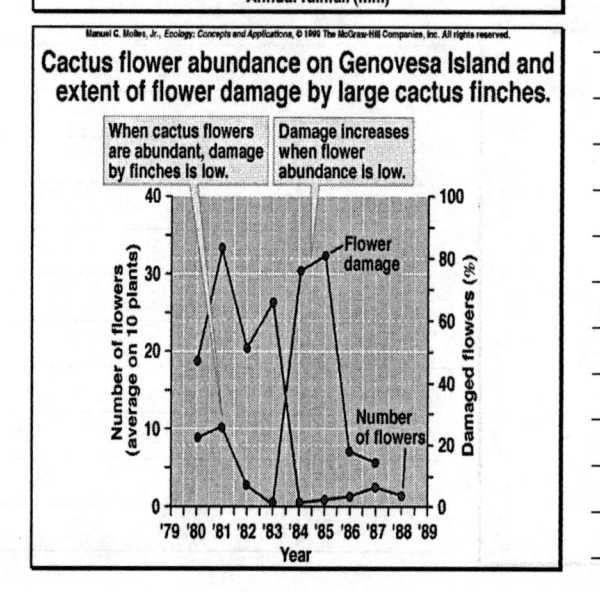

Cactus flower abundance on Genovesa Island and extent of flower damage by large cactus finches.

Figure 9.21

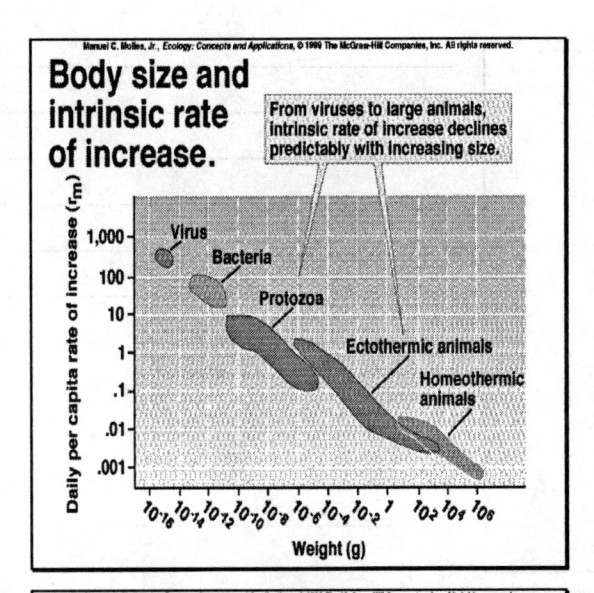

Body size and intrinsic rate of increase.

From viruses to large animals, intrinsic rate of increase declines predictably with increasing size.

Figure 9.22

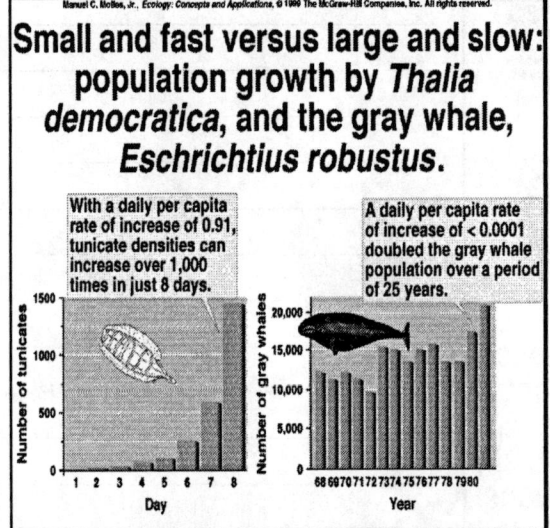

Small and fast versus large and slow: population growth by *Thalia democratica*, and the gray whale, *Eschrichtius robustus*.

With a daily per capita rate of increase of 0.91, tunicate densities can increase over 1,000 times in just 8 days.

A daily per capita rate of increase of < 0.0001 doubled the gray whale population over a period of 25 years.

Figure 9.23

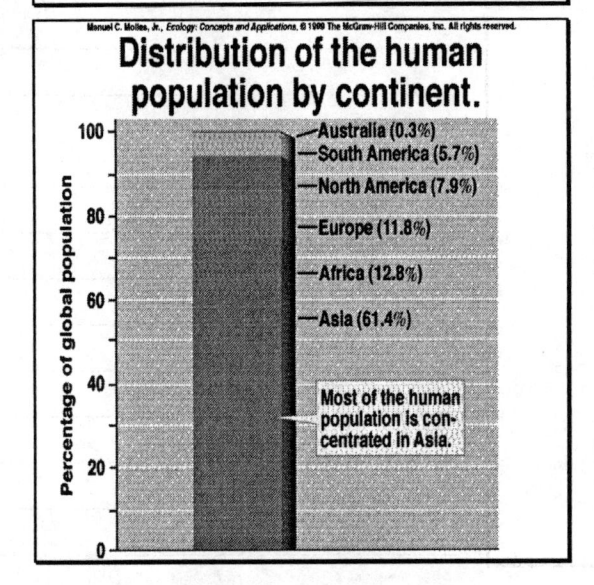

Distribution of the human population by continent.

- Australia (0.3%)
- South America (5.7%)
- North America (7.9%)
- Europe (11.8%)
- Africa (12.8%)
- Asia (61.4%)

Most of the human population is concentrated in Asia.

Figure 9.24

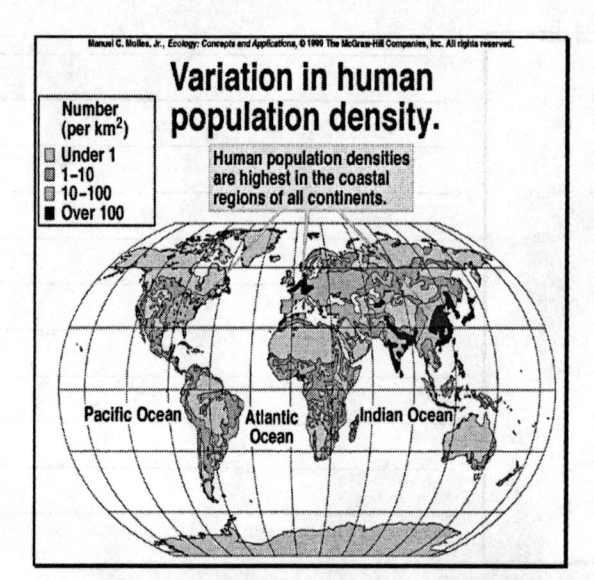

Figure 9.25

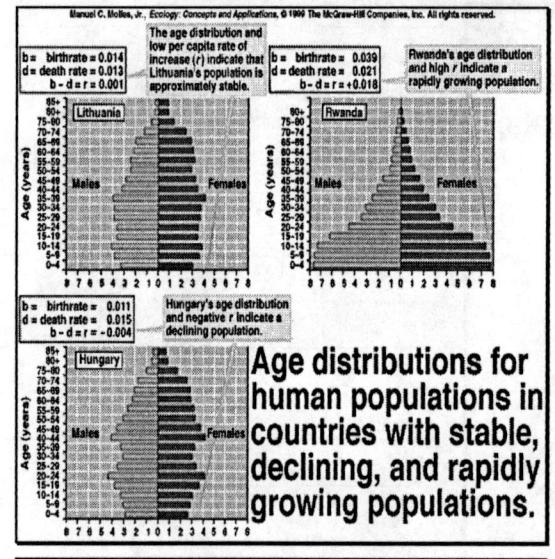

Figure 9.26

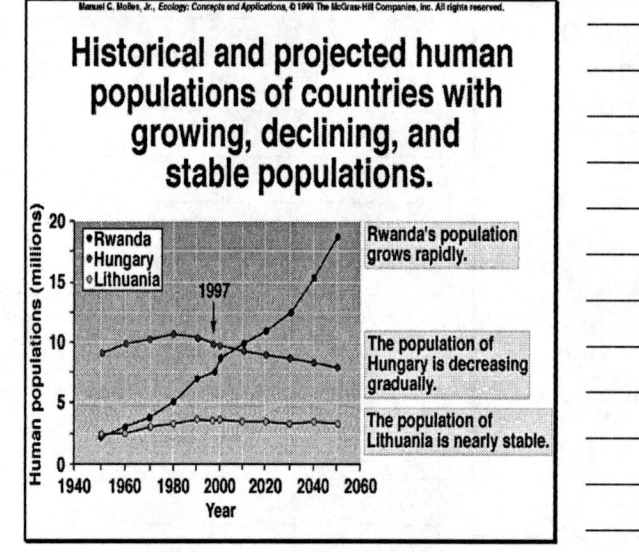

Figure 9.27

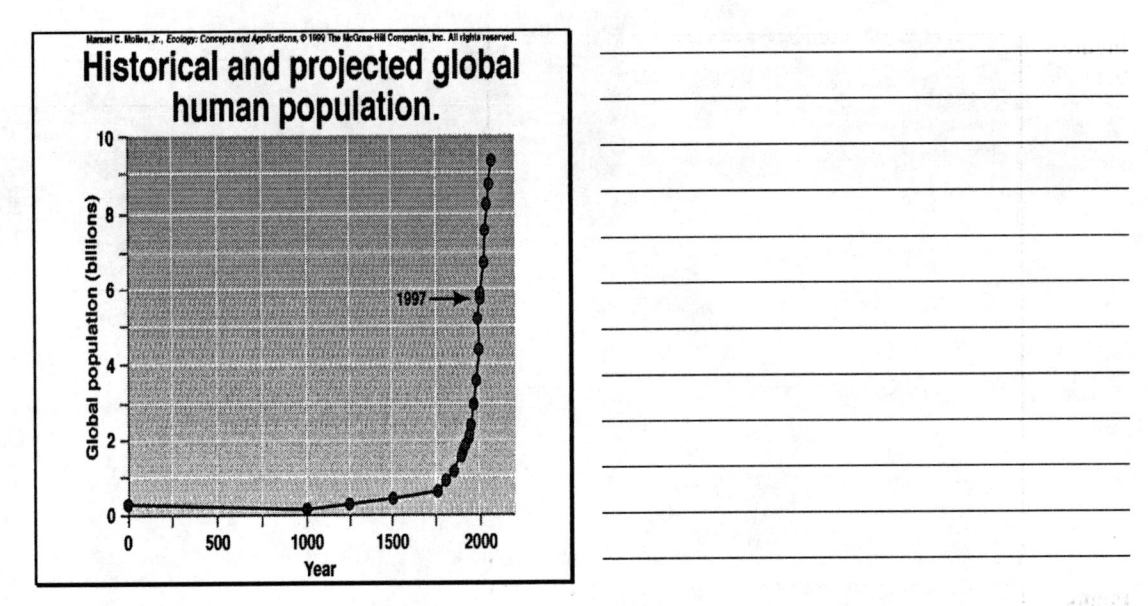

Historical and projected global human population.

Chapter 10

Ecology

Concepts and Applications
First Edition

Manuel C. Molles, Jr.

(c) The McGraw-Hill Companies, Inc.

Figure 10.3

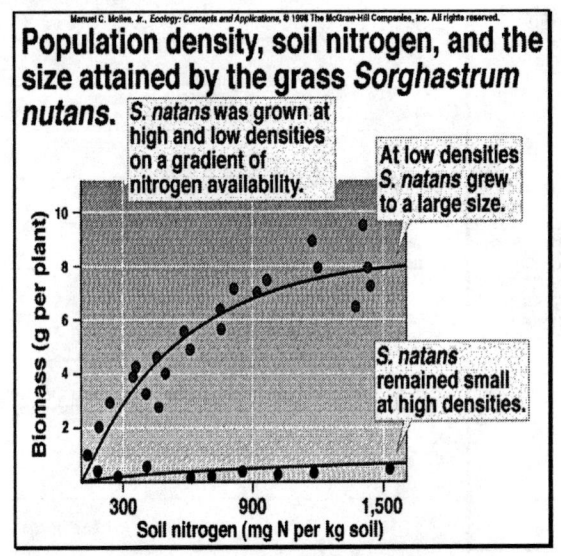

Population density, soil nitrogen, and the size attained by the grass *Sorghastrum nutans*.

S. natans was grown at high and low densities on a gradient of nitrogen availability.

At low densities S. natans grew to a large size.

S. natans remained small at high densities.

Figure 10.4

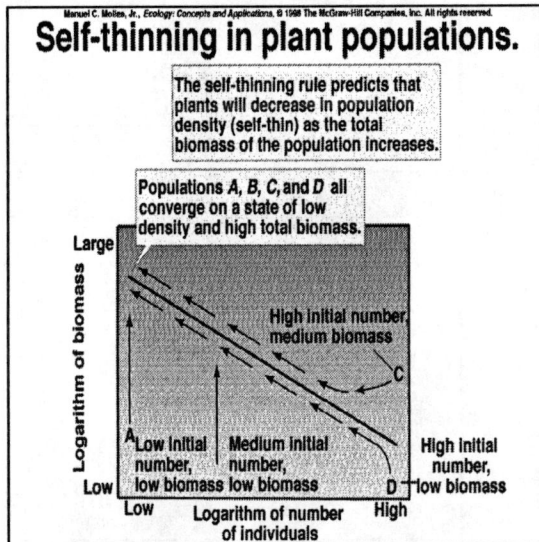

Self-thinning in plant populations.

The self-thinning rule predicts that plants will decrease in population density (self-thin) as the total biomass of the population increases.

Populations A, B, C, and D all converge on a state of low density and high total biomass.

High initial number, medium biomass

A - Low initial number, low biomass
Medium initial number, low biomass
D - High initial number, low biomass

Figure 10.5

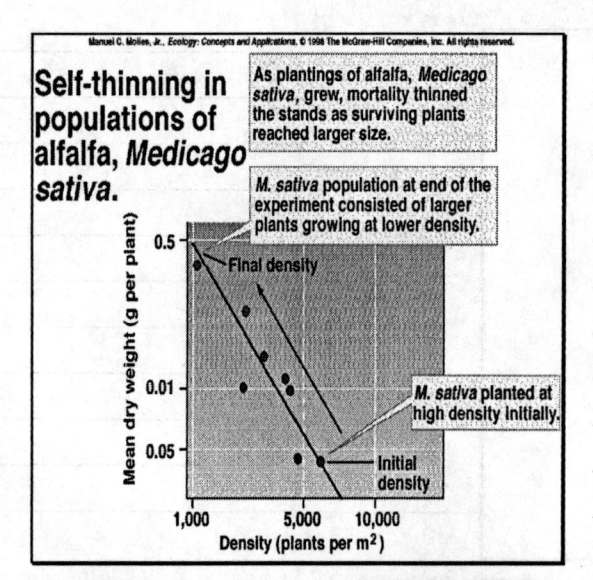

Self-thinning in populations of alfalfa, *Medicago sativa*.

As plantings of alfalfa, *Medicago sativa*, grew, mortality thinned the stands as surviving plants reached larger size.

M. sativa population at end of the experiment consisted of larger plants growing at lower density.

Final density

M. sativa planted at high density initially.

Initial density

Figure 10.6

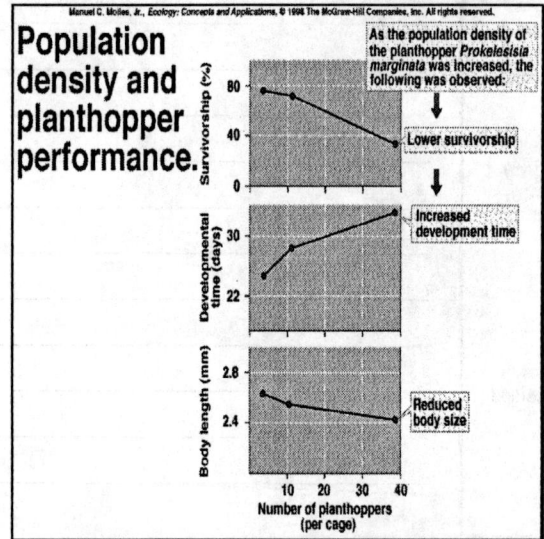

Population density and planthopper performance.

As the population density of the planthopper *Prokelesisia marginata* was increased, the following was observed:

Lower survivorship

Increased development time

Reduced body size

Figure 10.7

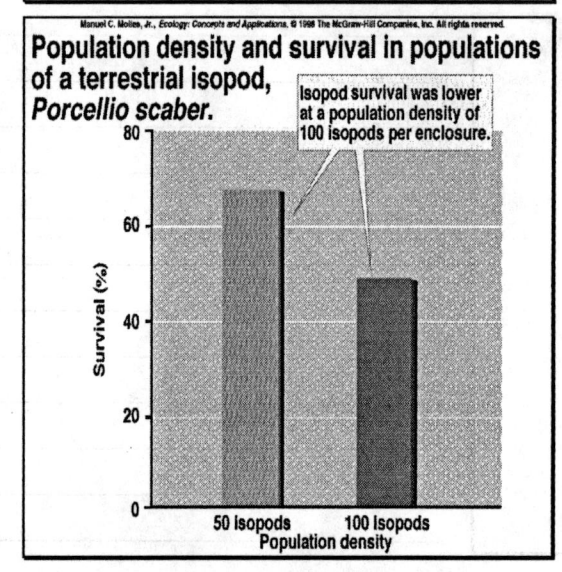

Population density and survival in populations of a terrestrial isopod, *Porcellio scaber*.

Isopod survival was lower at a population density of 100 isopods per enclosure.

Figure 10.8

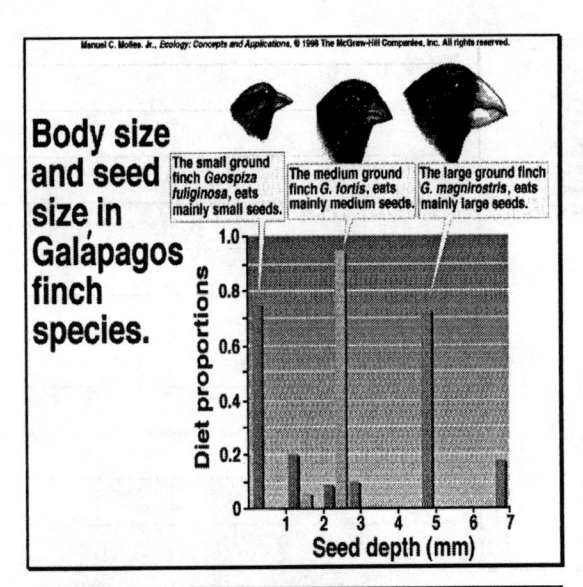

Body size and seed size in Galápagos finch species.

The small ground finch *Geospiza fuliginosa*, eats mainly small seeds.

The medium ground finch *G. fortis*, eats mainly medium seeds.

The large ground finch *G. magnirostris*, eats mainly large seeds.

Figure 10.9

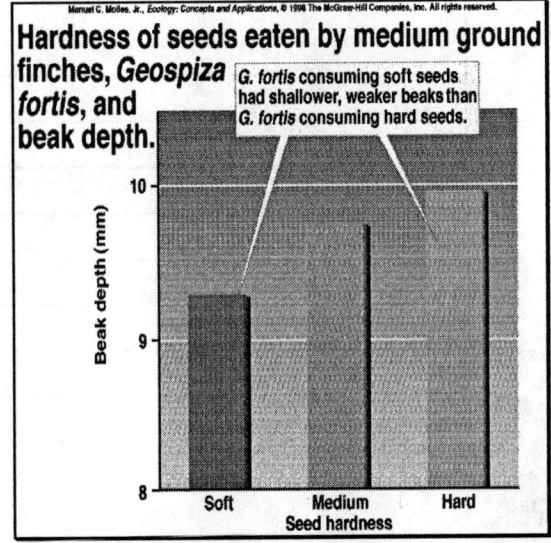

Hardness of seeds eaten by medium ground finches, *Geospiza fortis*, and beak depth.

G. fortis consuming soft seeds had shallower, weaker beaks than *G. fortis* consuming hard seeds.

Figure 10.10

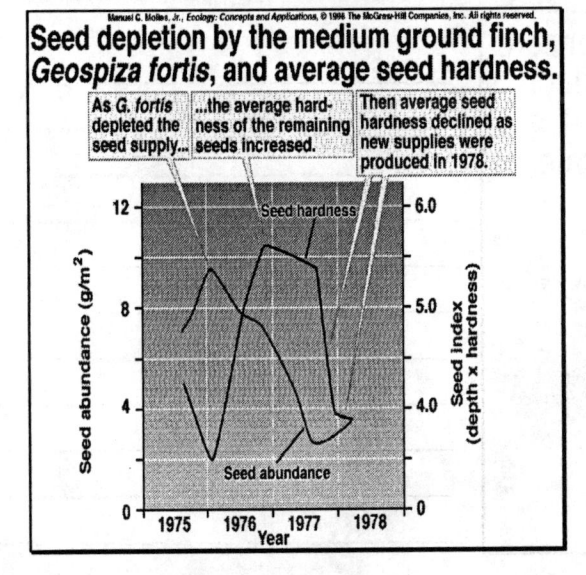

Seed depletion by the medium ground finch, *Geospiza fortis*, and average seed hardness.

As *G. fortis* depleted the seed supply...

...the average hardness of the remaining seeds increased.

Then average seed hardness declined as new supplies were produced in 1978.

Figure 10.11

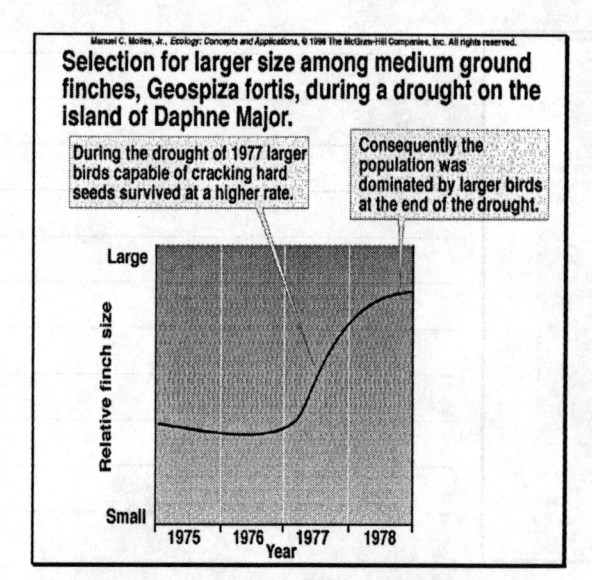

Selection for larger size among medium ground finches, Geospiza fortis, during a drought on the island of Daphne Major.

During the drought of 1977 larger birds capable of cracking hard seeds survived at a higher rate.

Consequently the population was dominated by larger birds at the end of the drought.

Figure 10.13

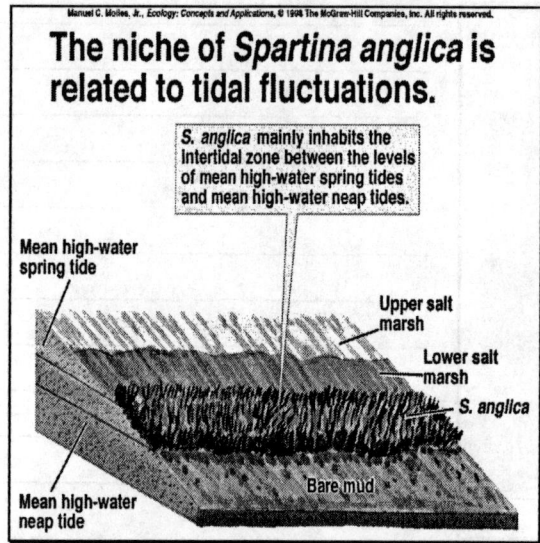

The niche of *Spartina anglica* is related to tidal fluctuations.

S. anglica mainly inhabits the intertidal zone between the levels of mean high-water spring tides and mean high-water neap tides.

Mean high-water spring tide
Upper salt marsh
Lower salt marsh
S. anglica
Bare mud
Mean high-water neap tide

Figure 10.14

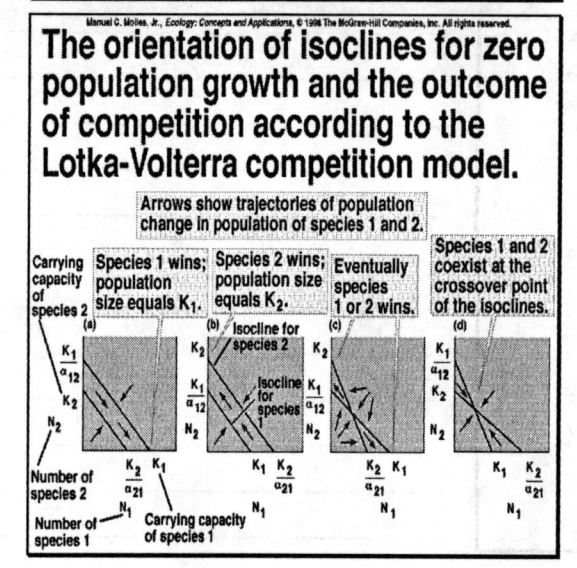

The orientation of isoclines for zero population growth and the outcome of competition according to the Lotka-Volterra competition model.

Arrows show trajectories of population change in population of species 1 and 2.

(a) Species 1 wins; population size equals K_1.
(b) Species 2 wins; population size equals K_2.
(c) Eventually species 1 or 2 wins.
(d) Species 1 and 2 coexist at the crossover point of the isoclines.

Figure 10.15

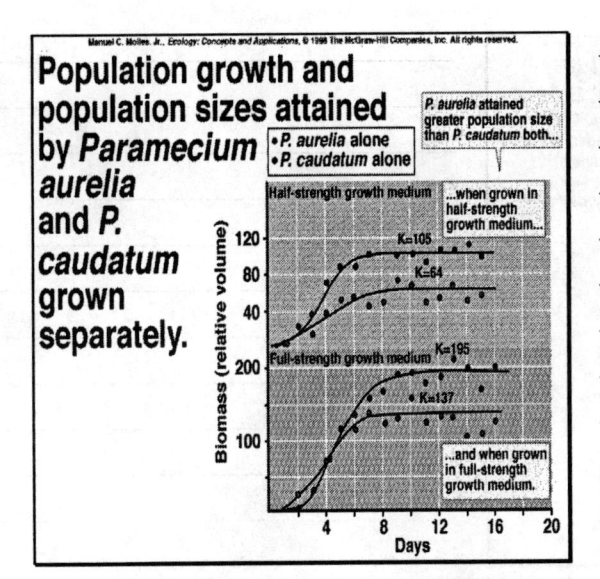

Population growth and population sizes attained by *Paramecium aurelia* and *P. caudatum* grown separately.

- *P. aurelia* alone
- *P. caudatum* alone

P. aurelia attained greater population size than *P. caudatum* both...

...when grown in half-strength growth medium...

...and when grown in full-strength growth medium.

Figure 10.16

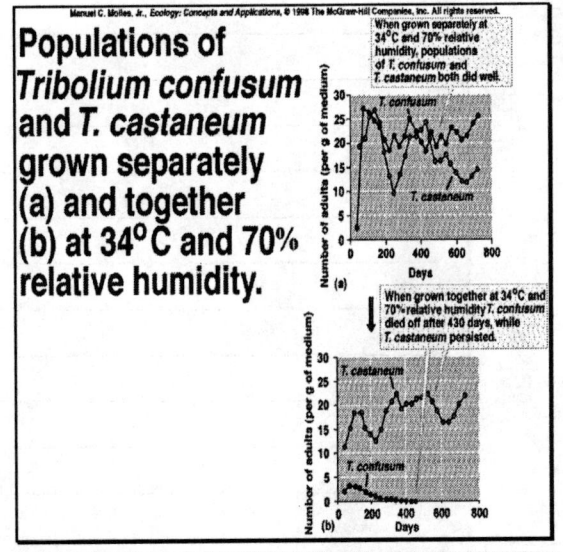

Populations of *Tribolium confusum* and *T. castaneum* grown separately (a) and together (b) at 34°C and 70% relative humidity.

When grown separately at 34°C and 70% relative humidity, populations of *T. confusum* and *T. castaneum* both did well.

When grown together at 34°C and 70% relative humidity *T. confusum* died off after 430 days, while *T. castaneum* persisted.

Figure 10.17

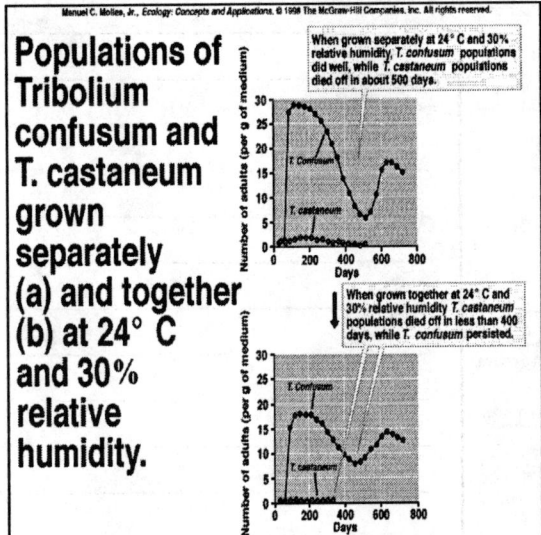

Populations of *Tribolium confusum* and *T. castaneum* grown separately (a) and together (b) at 24°C and 30% relative humidity.

When grown separately at 24°C and 30% relative humidity, *T. confusum* populations did well, while *T. castaneum* populations died off in about 500 days.

When grown together at 24°C and 30% relative humidity *T. castaneum* populations died off in less than 400 days, while *T. confusum* persisted.

Figure 10.19

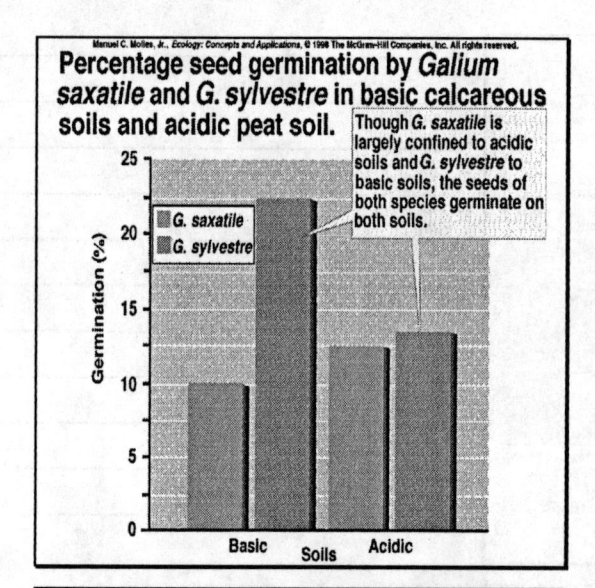

Percentage seed germination by *Galium saxatile* and *G. sylvestre* in basic calcareous soils and acidic peat soil.

Though *G. saxatile* is largely confined to acidic soils and *G. sylvestre* to basic soils, the seeds of both species germinate on both soils.

Figure 10.20

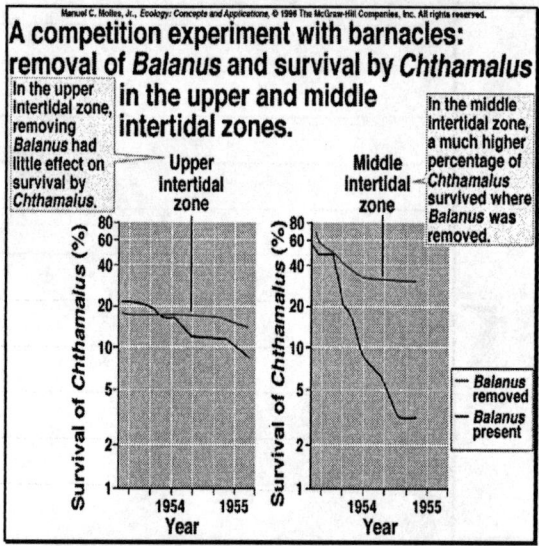

A competition experiment with barnacles: removal of *Balanus* and survival by *Chthamalus* in the upper and middle intertidal zones.

In the upper intertidal zone, removing *Balanus* had little effect on survival by *Chthamalus*.

In the middle intertidal zone, a much higher percentage of *Chthamalus* survived where *Balanus* was removed.

Figure 10.21

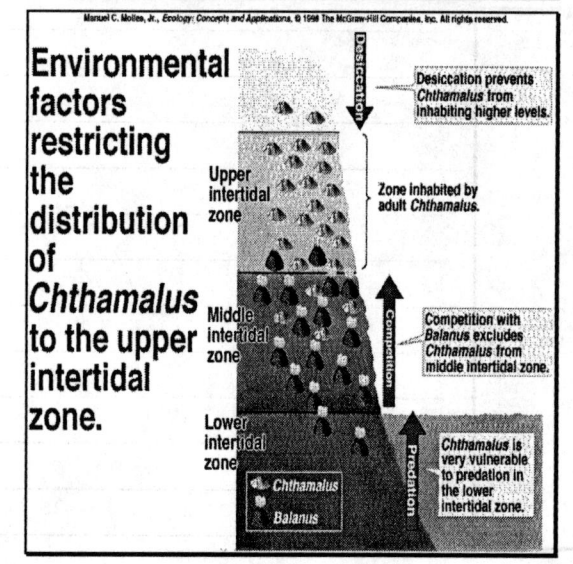

Environmental factors restricting the distribution of *Chthamalus* to the upper intertidal zone.

Desiccation prevents *Chthamalus* from inhabiting higher levels.

Zone inhabited by adult *Chthamalus*.

Competition with *Balanus* excludes *Chthamalus* from middle intertidal zone.

Chthamalus is very vulnerable to predation in the lower intertidal zone.

Figure 10.24: Responses by small granivorous and insectivorous rodents to removal of large granivorous *Dipodomys* species.

- *Dipodomys* numbers remained high on control plots throughout the study.
- Removal kept *Dipodomys* numbers at or near zero on the *Dipodomys* removal plots.
- In response, numbers of small granivorous rodents increased on the removal plots relative to the control plots.
- Meanwhile, numbers of insectivorous rodents did not differ on control and removal plots.

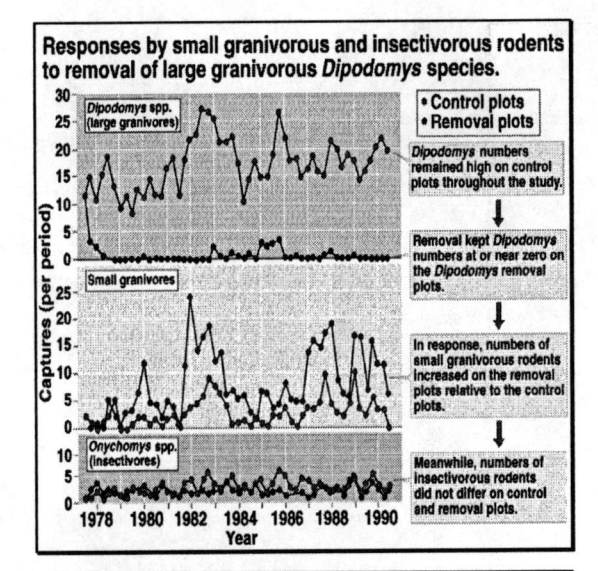

Figure 10.25: Responses of small granivorous and insectivorous rodents to a second removal experiment, which was preceded by several years of study before initiating *Dipodomys* removal.

- *Dipodomys* numbers were immediately reduced by the removal procedures on the removal plots.
- Small granivore numbers increased very quickly on the *Dipodomys* removal plots.
- Meanwhile, numbers of insectivorous rodents again did not change in response to *Dipodomys* removal.

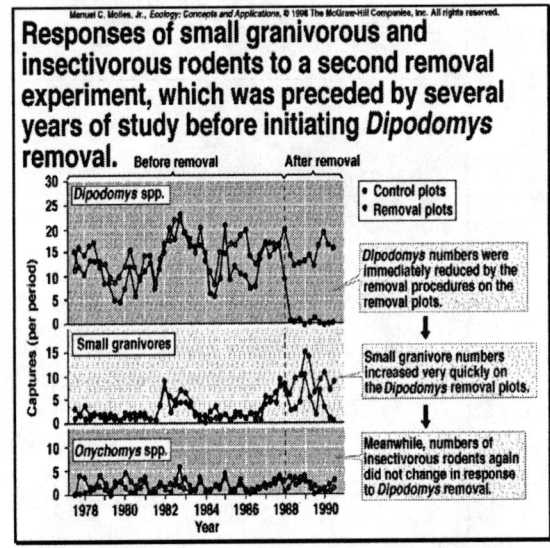

Figure 10.26: Evidence for character displacement in beak size in populations of the Galápagos finches *Geospiza fortis* and *G. fuliginosa*.

- Beak size distributions for *G. fortis* on Daphne Major.
- Compared to the population on Daphne Major, the beaks of *G. fortis* are significantly larger on the island of Santa Cruz, where it is sympatric with *G. fuliginosa*.
- Beak size distributions for sympatric populations of *G. fortis* and *G. fuliginosa* on Santa Cruz Island.
- Similarly, compared to the population on Los Hermanos, the beaks of *G. fuliginosa* are significantly smaller on Santa Cruz, where it is sympatric with *G. fortis*.
- Beak size distribution for *G. fuliginosa* on Los Hermanos.

89

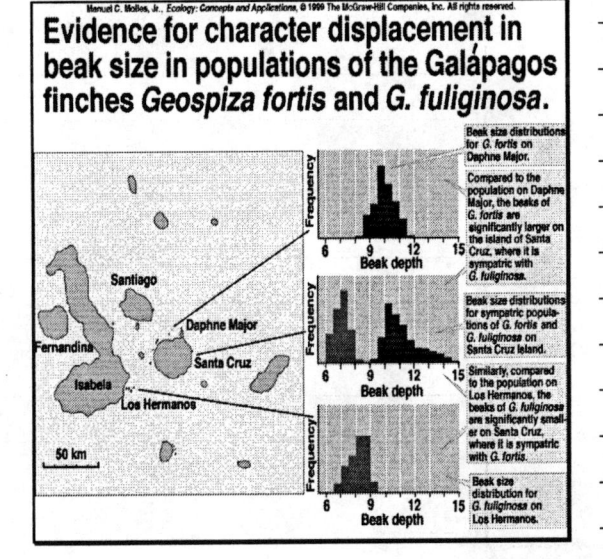

Chapter 11

Ecology
Concepts and Applications
First Edition

Manuel C. Molles, Jr.

(c) The McGraw-Hill Companies, Inc.

Figure 11.2

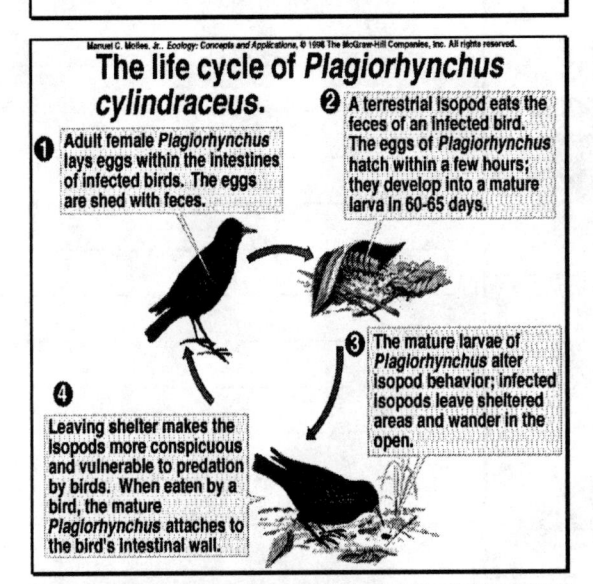

The life cycle of *Plagiorhynchus cylindraceus*.

❶ Adult female *Plagiorhynchus* lays eggs within the intestines of infected birds. The eggs are shed with feces.

❷ A terrestrial isopod eats the feces of an infected bird. The eggs of *Plagiorhynchus* hatch within a few hours; they develop into a mature larva in 60-65 days.

❸ The mature larvae of *Plagiorhynchus* alter isopod behavior; infected isopods leave sheltered areas and wander in the open.

❹ Leaving shelter makes the isopods more conspicuous and vulnerable to predation by birds. When eaten by a bird, the mature *Plagiorhynchus* attaches to the bird's intestinal wall.

Figure 11.3

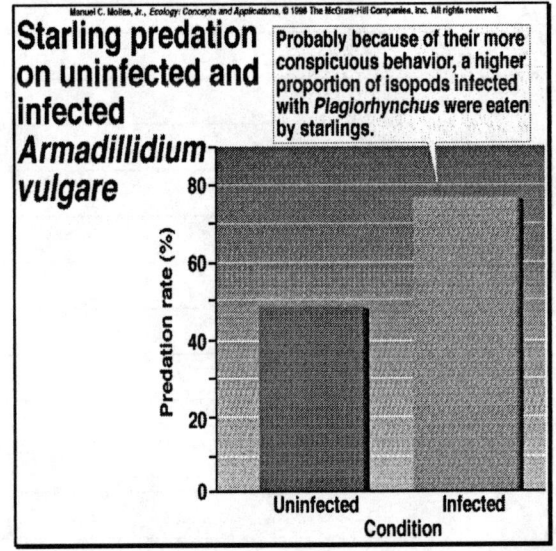

Starling predation on uninfected and infected *Armadillidium vulgare*

Probably because of their more conspicuous behavior, a higher proportion of isopods infected with *Plagiorhynchus* were eaten by starlings.

Figure 11.6

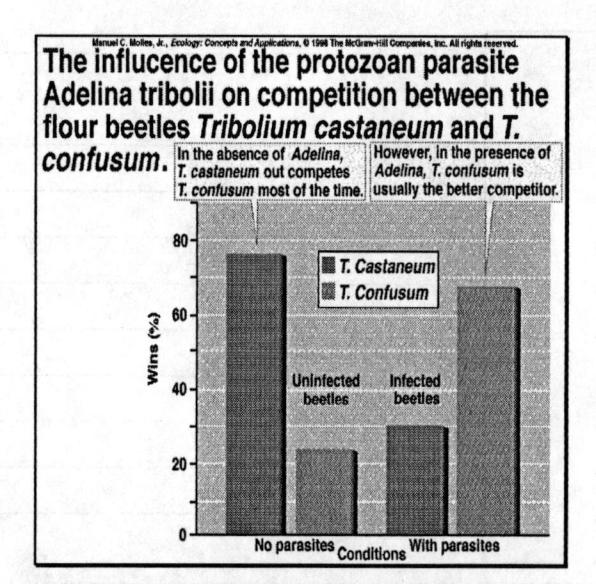

The influence of the protozoan parasite Adelina tribolii on competition between the flour beetles *Tribolium castaneum* and *T. confusum*.

In the absence of *Adelina*, *T. castaneum* out competes *T. confusum* most of the time.

However, in the presence of *Adelina*, *T. confusum* is usually the better competitor.

Figure 11.7

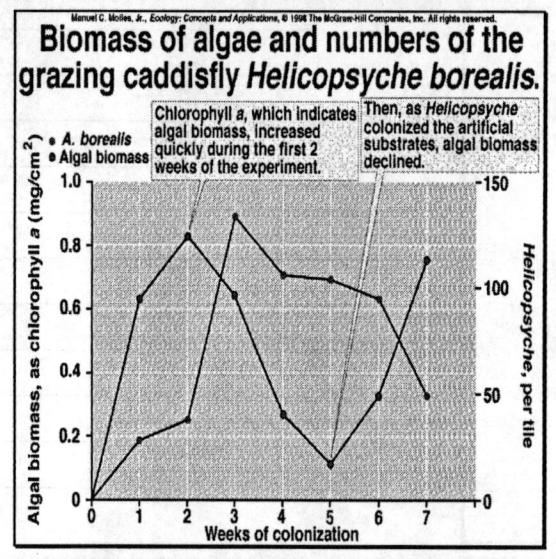

Biomass of algae and numbers of the grazing caddisfly *Helicopsyche borealis*.

Chlorophyll a, which indicates algal biomass, increased quickly during the first 2 weeks of the experiment.

Then, as *Helicopsyche* colonized the artificial substrates, algal biomass declined.

Figure 11.8

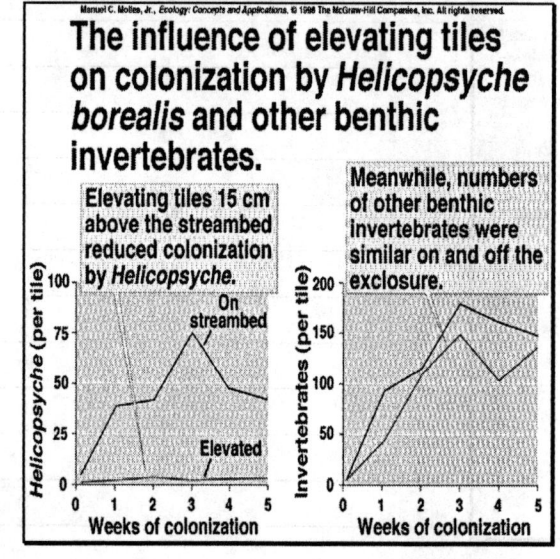

The influence of elevating tiles on colonization by *Helicopsyche borealis* and other benthic invertebrates.

Elevating tiles 15 cm above the streambed reduced colonization by *Helicopsyche*.

Meanwhile, numbers of other benthic invertebrates were similar on and off the exclosure.

Figure 11.9: **Influence of excluding *Helicopsyche borealis* on abundance of bacteria and algae.**

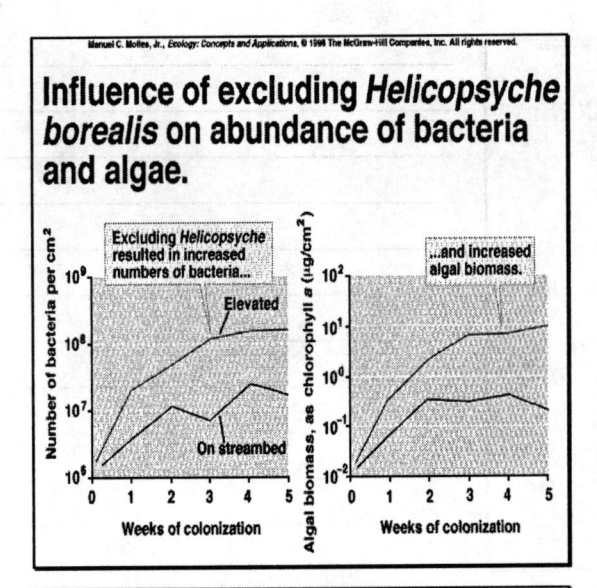

Figure 11.12: **The spread of mange in red foxes across Sweden.**

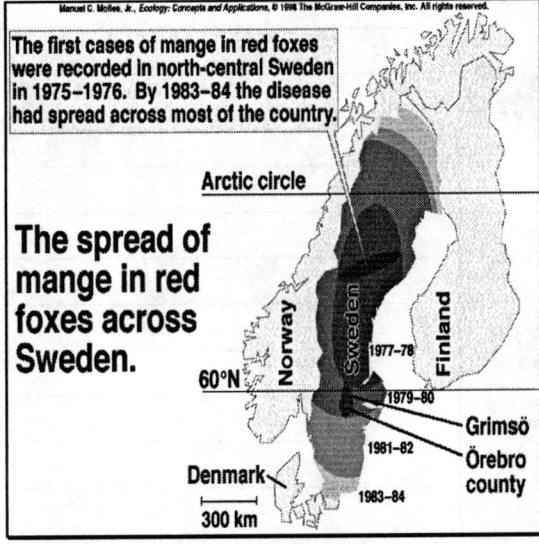

Figure 11.13: **The numbers of foxes and mountain hares in five counties in Sweden estimated from hunters' harvest records.**

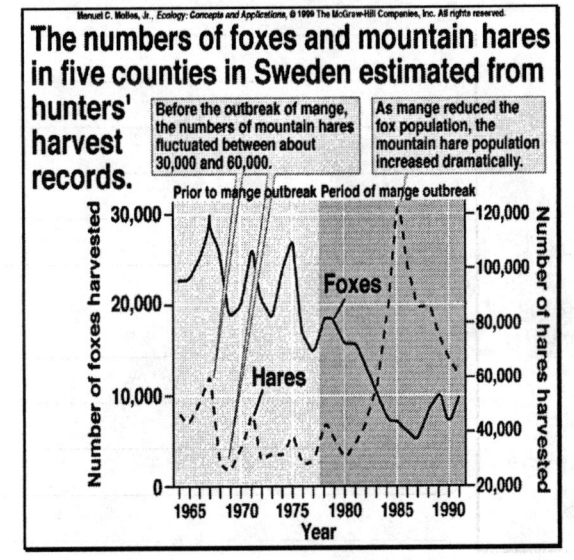

Figure 11.14

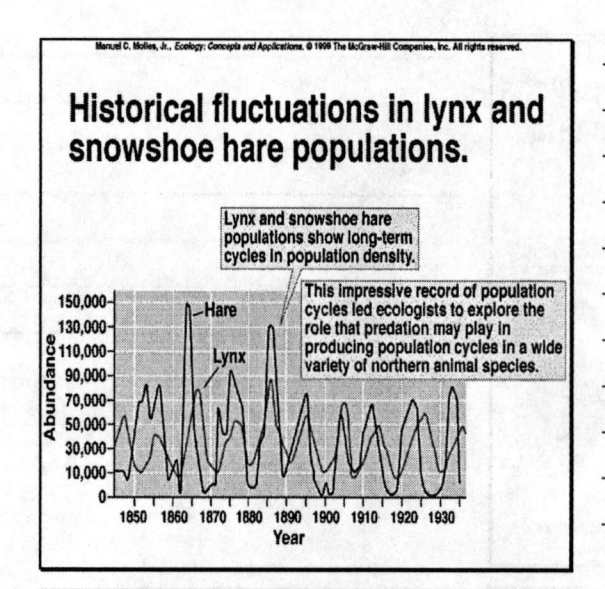

Historical fluctuations in lynx and snowshoe hare populations.

Lynx and snowshoe hare populations show long-term cycles in population density.

This impressive record of population cycles led ecologists to explore the role that predation may play in producing population cycles in a wide variety of northern animal species.

Figure 11.15

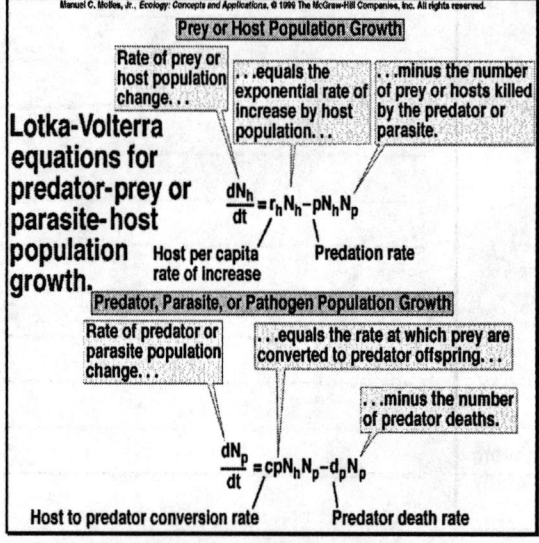

Lotka-Volterra equations for predator-prey or parasite-host population growth.

Prey or Host Population Growth

Rate of prey or host population change... equals the exponential rate of increase by host population... minus the number of prey or hosts killed by the predator or parasite.

$$\frac{dN_h}{dt} = r_h N_h - p N_h N_p$$

Host per capita rate of increase — Predation rate

Predator, Parasite, or Pathogen Population Growth

Rate of predator or parasite population change... equals the rate at which prey are converted to predator offspring... minus the number of predator deaths.

$$\frac{dN_p}{dt} = c p N_h N_p - d_p N_p$$

Host to predator conversion rate — Predator death rate

Figure 11.16

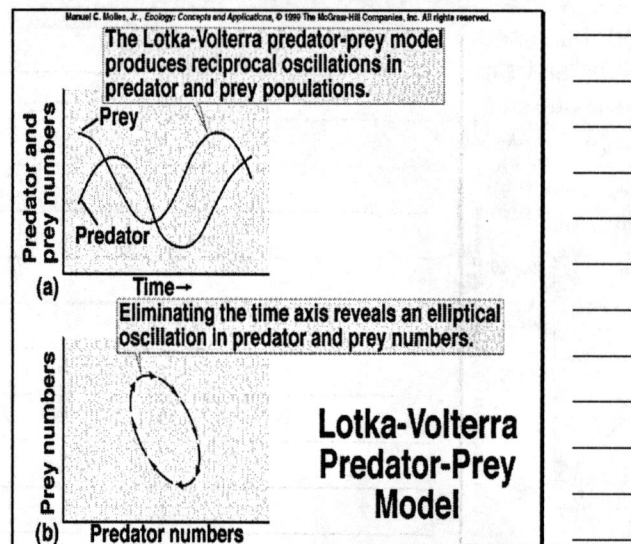

The Lotka-Volterra predator-prey model produces reciprocal oscillations in predator and prey populations.

(a) Predator and prey numbers vs. Time

Eliminating the time axis reveals an elliptical oscillation in predator and prey numbers.

(b) Prey numbers vs. Predator numbers

Lotka-Volterra Predator-Prey Model

Figure 11.17

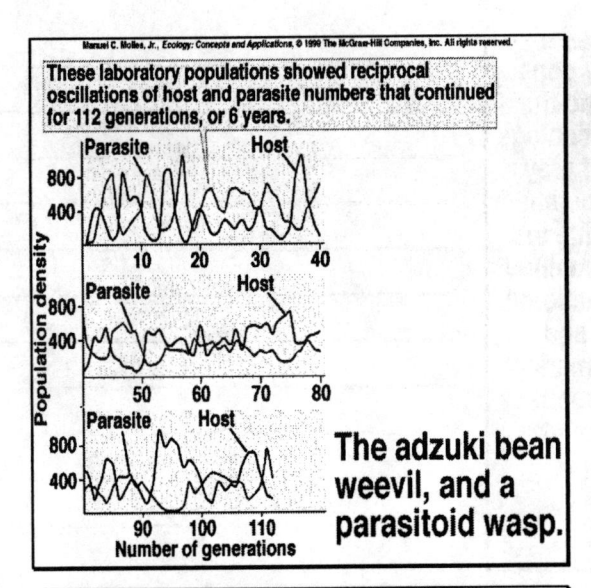

These laboratory populations showed reciprocal oscillations of host and parasite numbers that continued for 112 generations, or 6 years.

The adzuki bean weevil, and a parasitoid wasp.

Figure 11.18

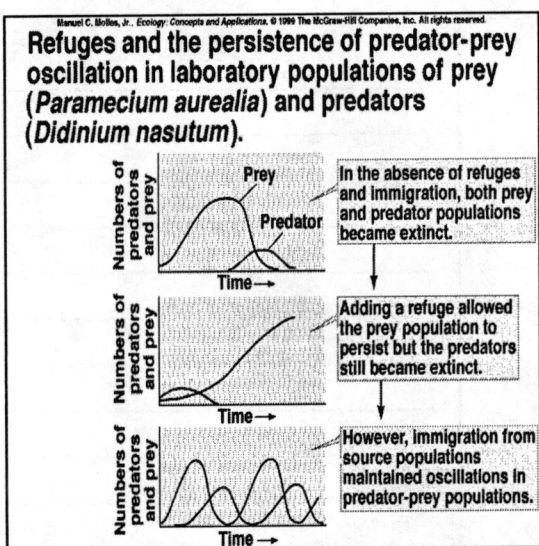

Refuges and the persistence of predator-prey oscillation in laboratory populations of prey (*Paramecium aurealia*) and predators (*Didinium nasutum*).

In the absence of refuges and immigration, both prey and predator populations became extinct.

Adding a refuge allowed the prey population to persist but the predators still became extinct.

However, immigration from source populations maintained oscillations in predator-prey populations.

Figure 11.19

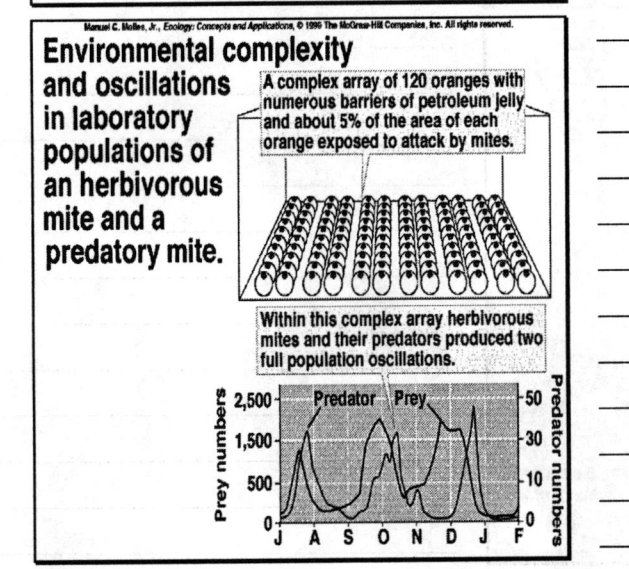

Environmental complexity and oscillations in laboratory populations of an herbivorous mite and a predatory mite.

A complex array of 120 oranges with numerous barriers of petroleum jelly and about 5% of the area of each orange exposed to attack by mites.

Within this complex array herbivorous mites and their predators produced two full population oscillations.

Figure 11.20

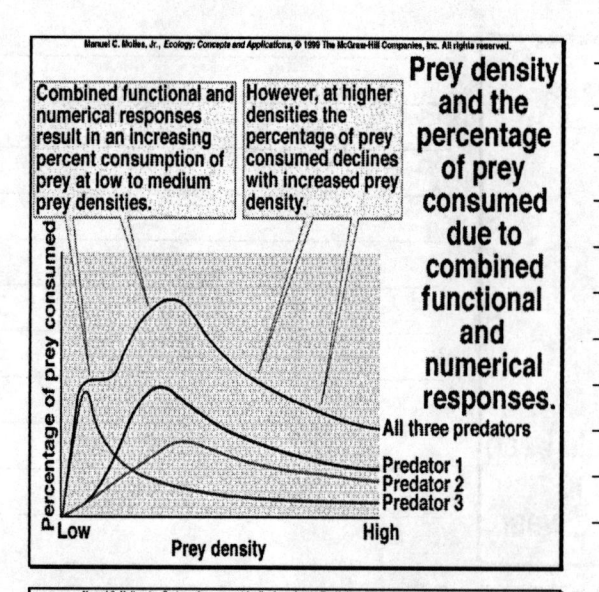

Prey density and the percentage of prey consumed due to combined functional and numerical responses.

Combined functional and numerical responses result in an increasing percent consumption of prey at low to medium prey densities.

However, at higher densities the percentage of prey consumed declines with increased prey density.

- All three predators
- Predator 1
- Predator 2
- Predator 3

Y-axis: Percentage of prey consumed
X-axis: Prey density (Low to High)

Figure 11.21

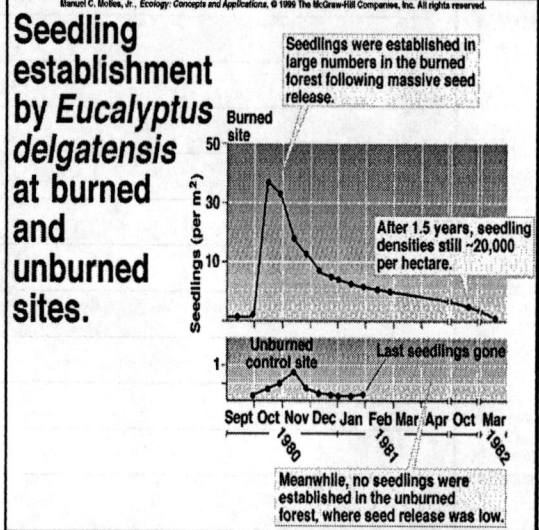

Seedling establishment by *Eucalyptus delgatensis* at burned and unburned sites.

Seedlings were established in large numbers in the burned forest following massive seed release.

After 1.5 years, seedling densities still ~20,000 per hectare.

Burned site
Seedlings (per m^2): 50, 30, 10, 1

Unburned control site

Last seedlings gone

Sept Oct Nov Dec Jan Feb Mar Apr Oct Mar
1990   1991   1992

Meanwhile, no seedlings were established in the unburned forest, where seed release was low.

Figure 11.22

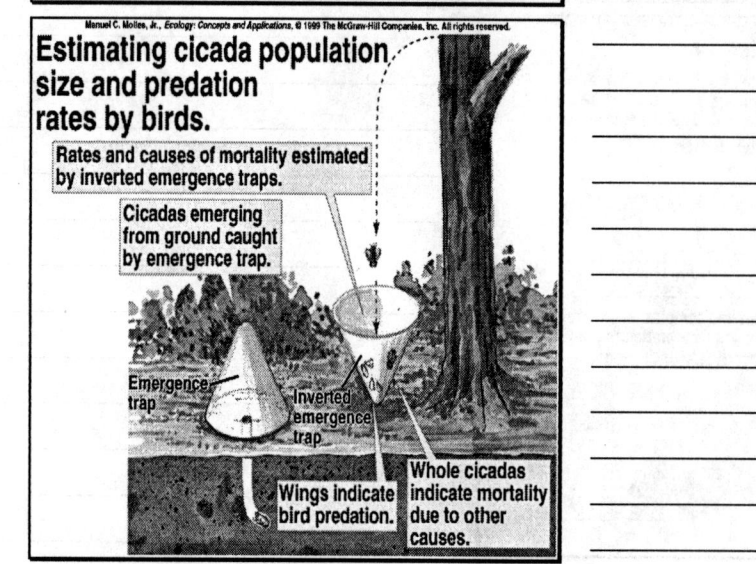

Estimating cicada population size and predation rates by birds.

Rates and causes of mortality estimated by inverted emergence traps.

Cicadas emerging from ground caught by emergence trap.

Emergence trap

Inverted emergence trap

Wings indicate bird predation.

Whole cicadas indicate mortality due to other causes.

Figure 11.23

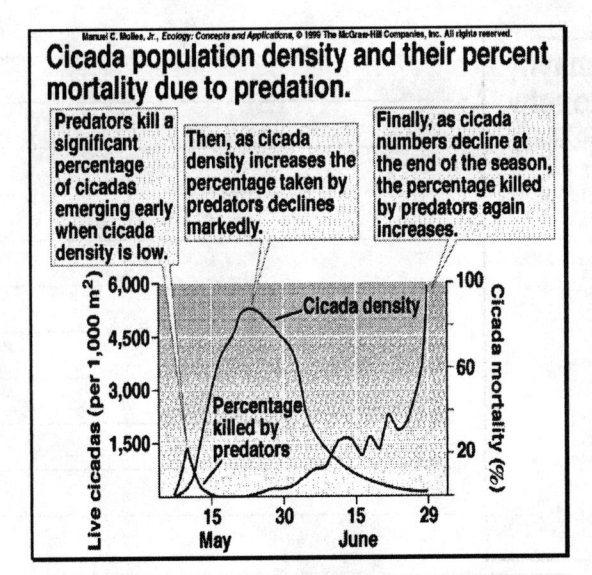

Figure 11.25

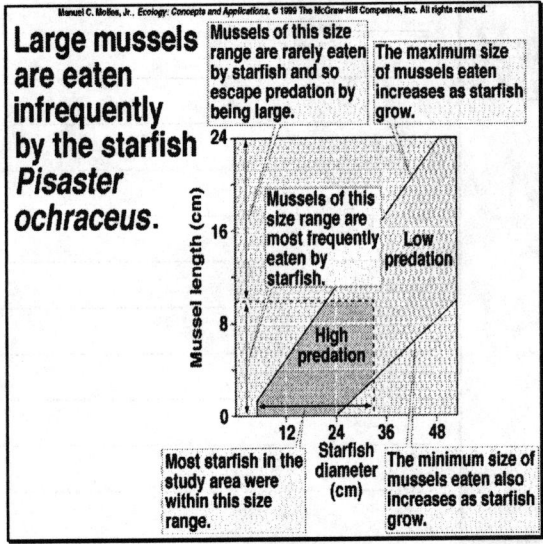

Figure 11.26

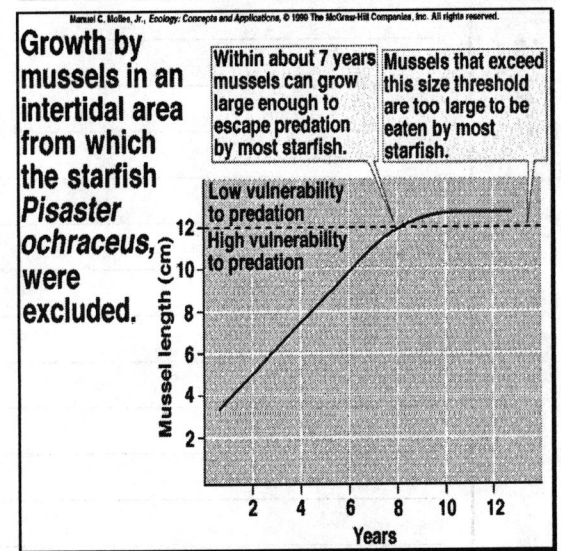

Figure 11.27

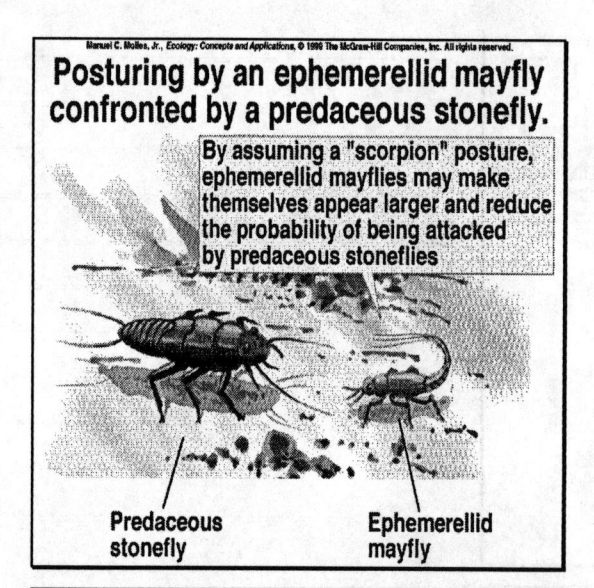

Posturing by an ephemerellid mayfly confronted by a predaceous stonefly.

By assuming a "scorpion" posture, ephemerellid mayflies may make themselves appear larger and reduce the probability of being attacked by predaceous stoneflies

Predaceous stonefly

Ephemerellid mayfly

Figure 11.28

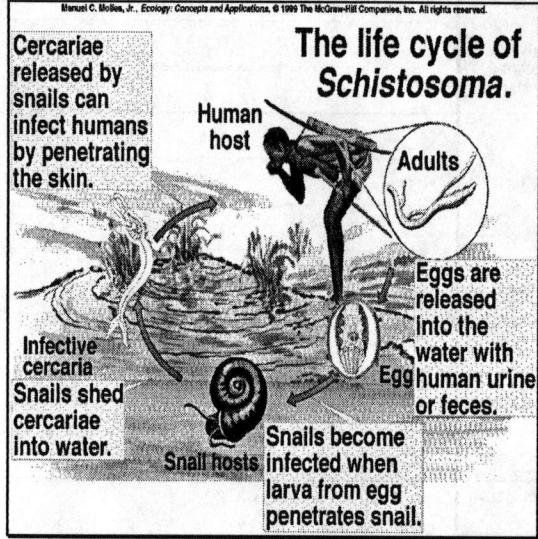

The life cycle of *Schistosoma*.

Cercariae released by snails can infect humans by penetrating the skin.

Human host

Adults

Eggs are released into the water with human urine or feces.

Egg

Infective cercaria Snails shed cercariae into water.

Snail hosts

Snails become infected when larva from egg penetrates snail.

Figure 11.29

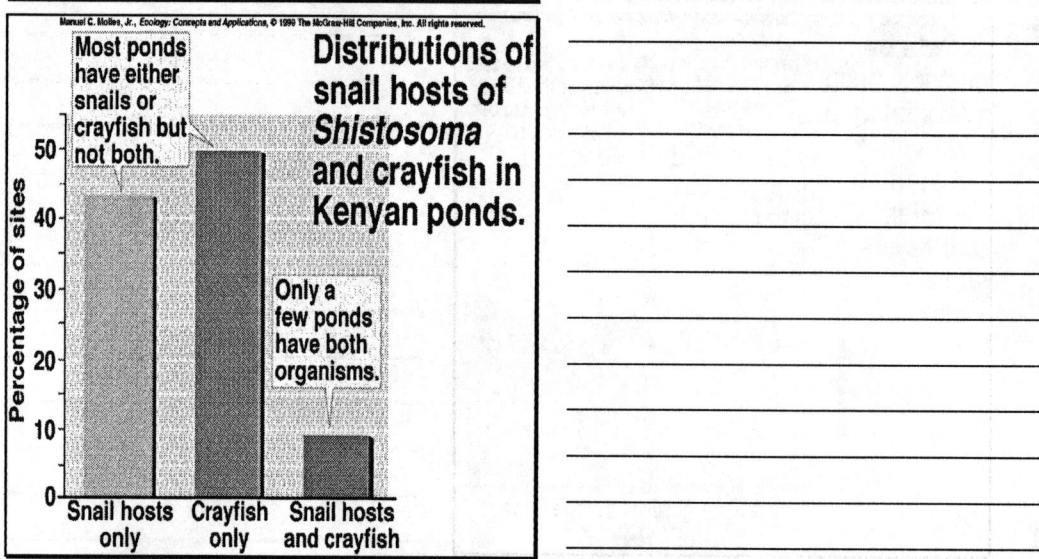

Distributions of snail hosts of *Shistosoma* and crayfish in Kenyan ponds.

Most ponds have either snails or crayfish but not both.

Only a few ponds have both organisms.

Snail hosts only | Crayfish only | Snail hosts and crayfish

Percentage of sites

Chapter 12

Ecology

Concepts and Applications
First Edition

Manuel C. Molles, Jr.

(c) The McGraw-Hill Companies, Inc.

Figure 12.3

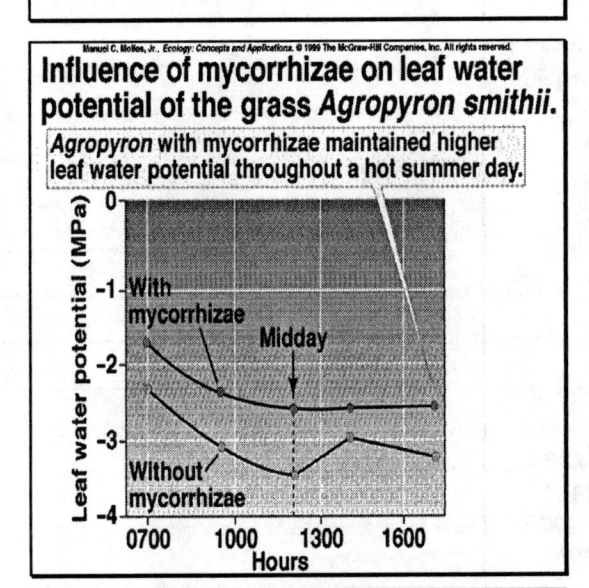

Influence of mycorrhizae on leaf water potential of the grass *Agropyron smithii*.

Agropyron with mycorrhizae maintained higher leaf water potential throughout a hot summer day.

Figure 12.4

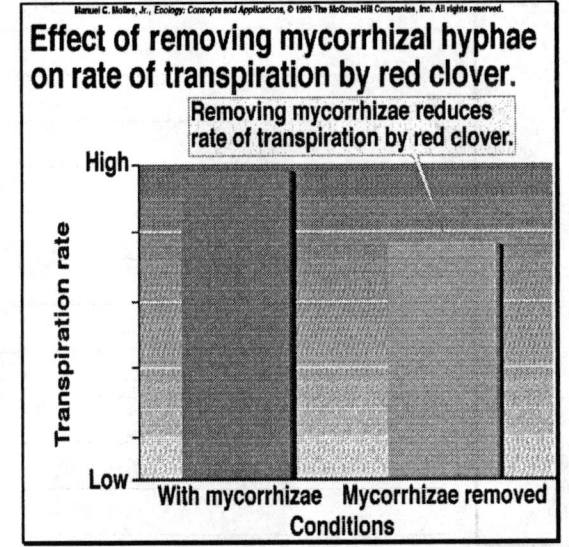

Effect of removing mycorrhizal hyphae on rate of transpiration by red clover.

Removing mycorrhizae reduces rate of transpiration by red clover.

Figure 12.5

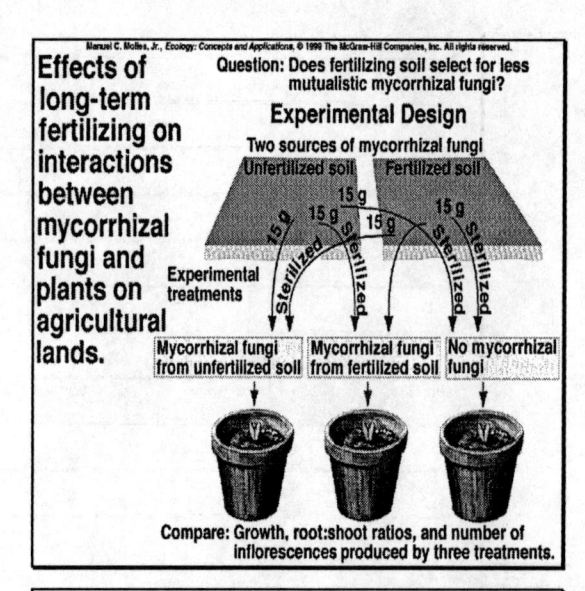

Figure 12.6

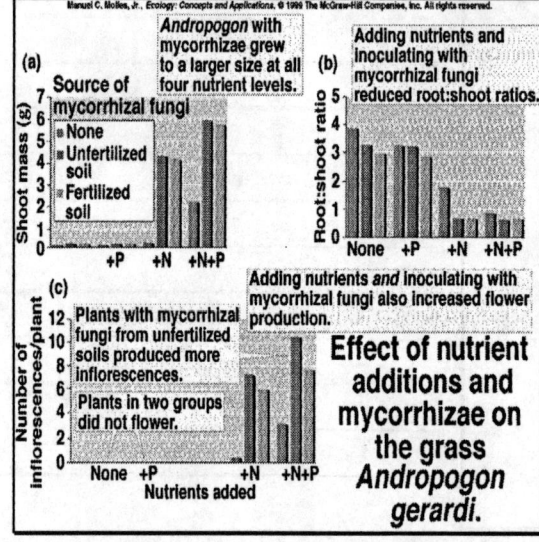

Figure 12.8

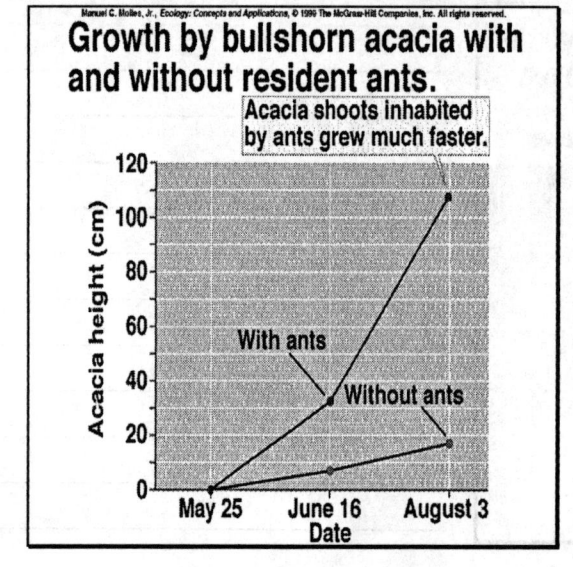

Figure 12.9

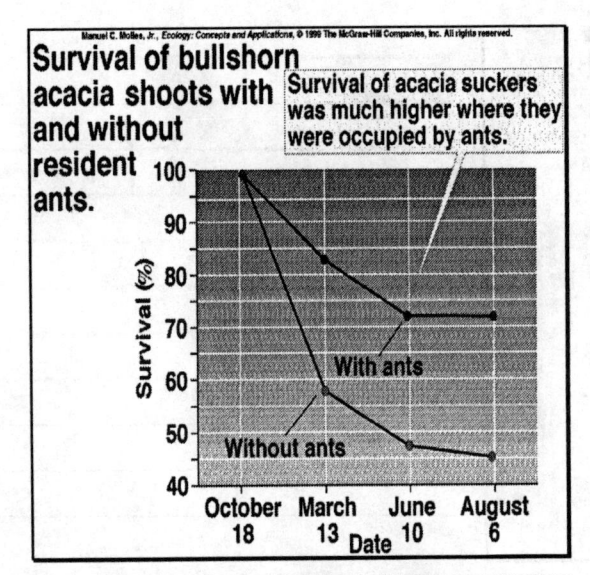

Survival of bullshorn acacia shoots with and without resident ants.

Survival of acacia suckers was much higher where they were occupied by ants.

[Graph: Survival (%) vs Date (October 18, March 13, June 10, August 6); "With ants" line declines from 100 to ~72; "Without ants" line declines from 100 to ~45]

Figure 12.10

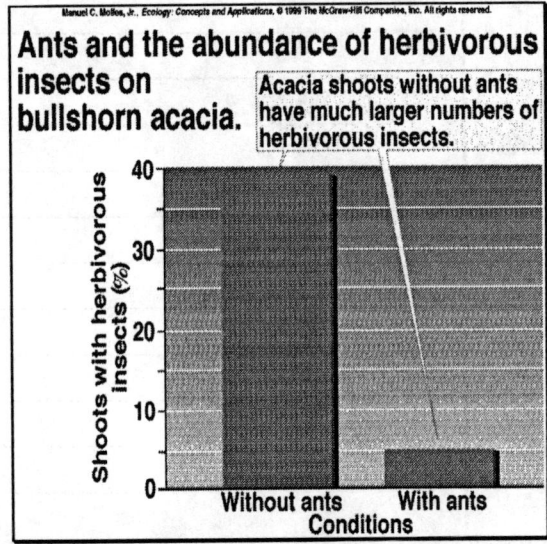

Ants and the abundance of herbivorous insects on bullshorn acacia.

Acacia shoots without ants have much larger numbers of herbivorous insects.

[Bar graph: Shoots with herbivorous insects (%) — Without ants ~40, With ants ~5]

Figure 12.11

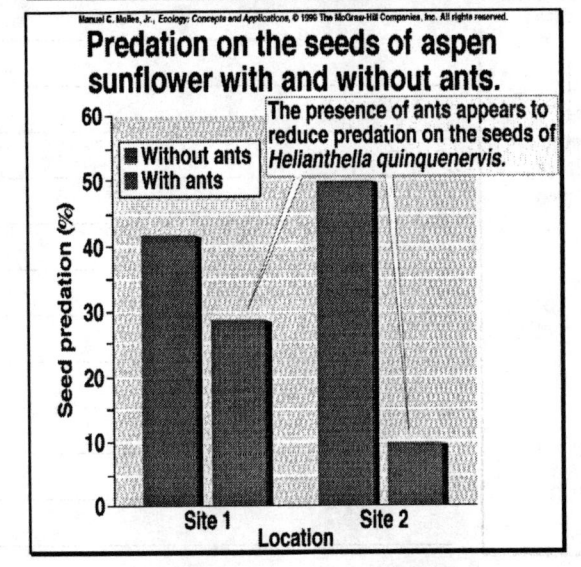

Predation on the seeds of aspen sunflower with and without ants.

The presence of ants appears to reduce predation on the seeds of *Helianthella quinquenervis*.

[Bar graph: Seed predation (%) at Site 1 and Site 2, comparing Without ants and With ants]

Figure 12.12

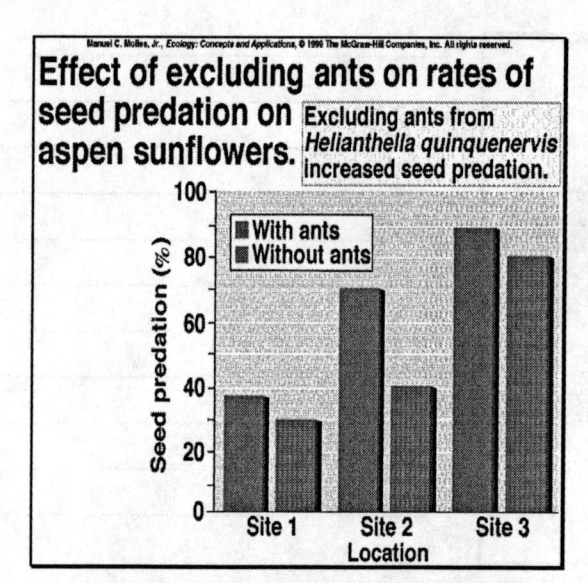

Effect of excluding ants on rates of seed predation on aspen sunflowers.

Excluding ants from *Helianthella quinquenervis* increased seed predation.

Figure 12.13

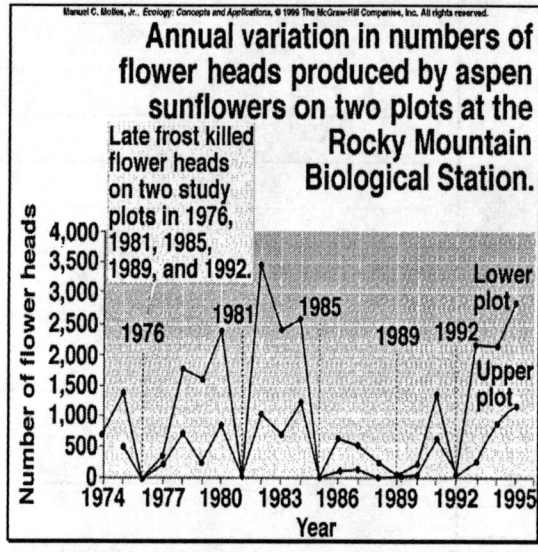

Annual variation in numbers of flower heads produced by aspen sunflowers on two plots at the Rocky Mountain Biological Station.

Late frost killed flower heads on two study plots in 1976, 1981, 1985, 1989, and 1992.

Figure 12.14

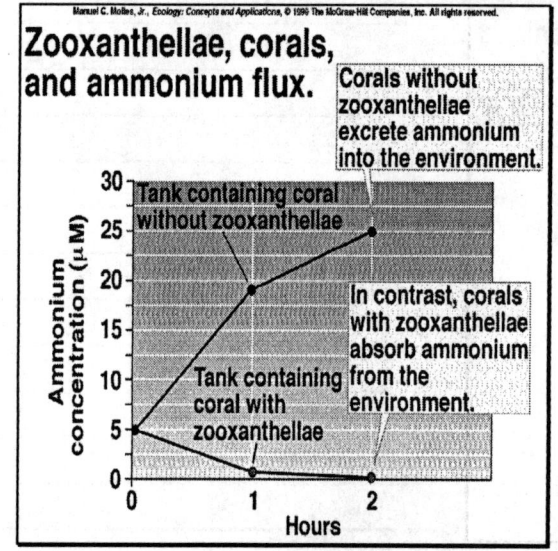

Zooxanthellae, corals, and ammonium flux.

Corals without zooxanthellae excrete ammonium into the environment.

In contrast, corals with zooxanthellae absorb ammonium from the environment.

Figure 12.16

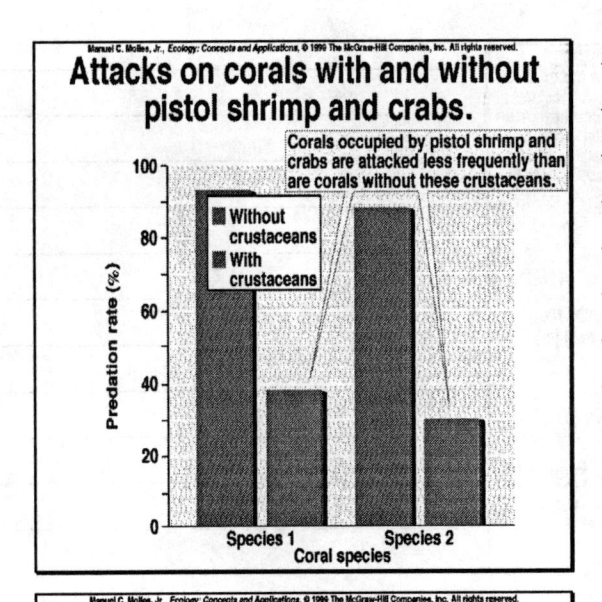

Attacks on corals with and without pistol shrimp and crabs.

Corals occupied by pistol shrimp and crabs are attacked less frequently than are corals without these crustaceans.

- Without crustaceans
- With crustaceans

Predation rate (%) vs Coral species (Species 1, Species 2)

Figure 12.17

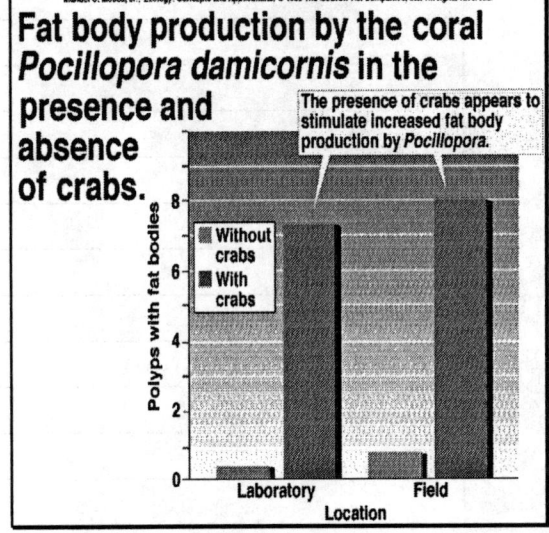

Fat body production by the coral *Pocillopora damicornis* in the presence and absence of crabs.

The presence of crabs appears to stimulate increased fat body production by *Pocillopora*.

- Without crabs
- With crabs

Polyps with fat bodies vs Location (Laboratory, Field)

Figure 12.20

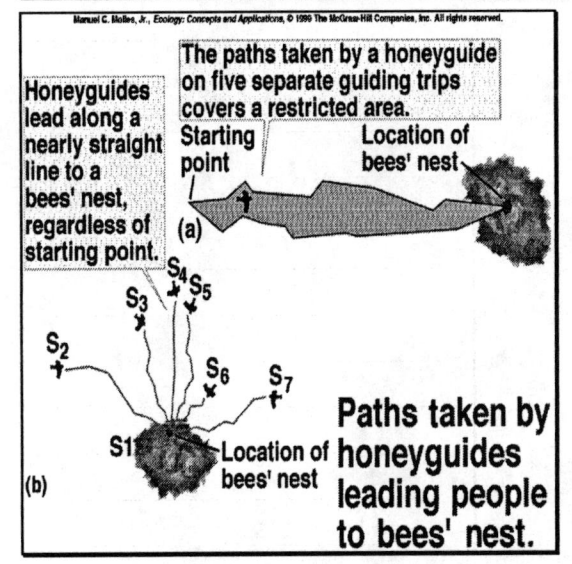

Honeyguides lead along a nearly straight line to a bees' nest, regardless of starting point.

The paths taken by a honeyguide on five separate guiding trips covers a restricted area.

(a) Starting point → Location of bees' nest

(b) S1, S2, S3, S4, S5, S6, S7 → Location of bees' nest

Paths taken by honeyguides leading people to bees' nest.

Figure 12.21

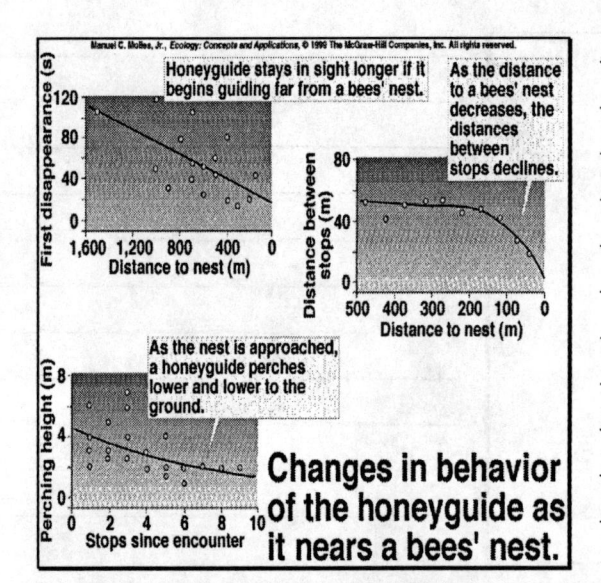

Changes in behavior of the honeyguide as it nears a bees' nest.

Honeyguide stays in sight longer if it begins guiding far from a bees' nest.

As the distance to a bees' nest decreases, the distances between stops declines.

As the nest is approached, a honeyguide perches lower and lower to the ground.

Figure 12.22

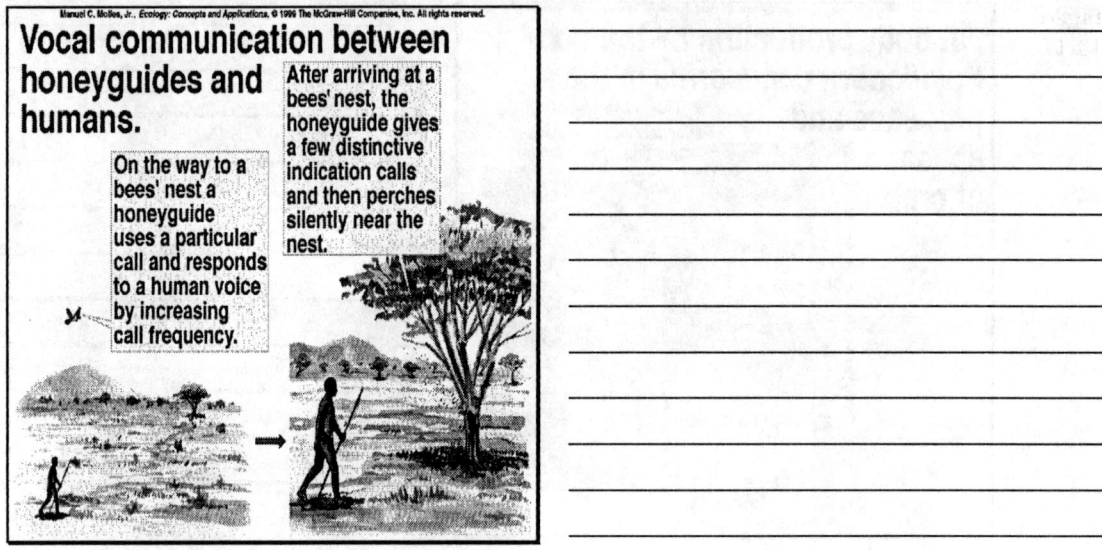

Vocal communication between honeyguides and humans.

On the way to a bees' nest a honeyguide uses a particular call and responds to a human voice by increasing call frequency.

After arriving at a bees' nest, the honeyguide gives a few distinctive indication calls and then perches silently near the nest.

Chapter 13

Ecology
Concepts and Applications
First Edition

Manuel C. Molles, Jr.

(c) The McGraw-Hill Companies, Inc.

Figure 13.3

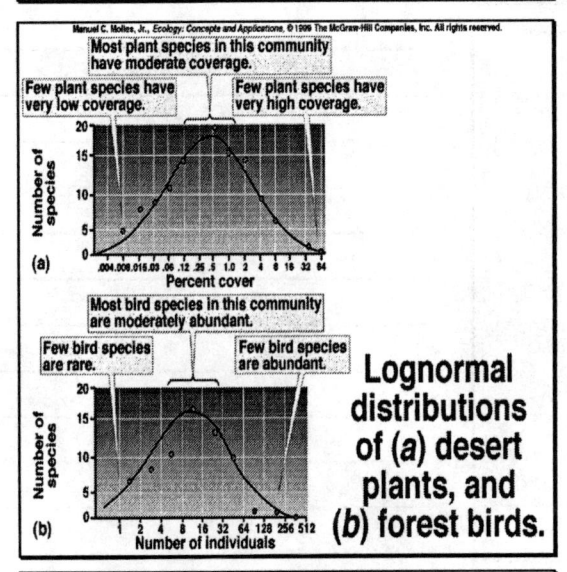

Lognormal distributions of (*a*) desert plants, and (*b*) forest birds.

Figure 13.4

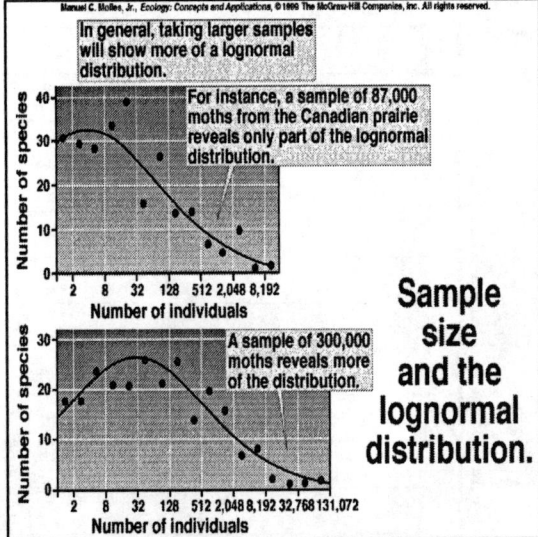

Sample size and the lognormal distribution.

Figure 13.5

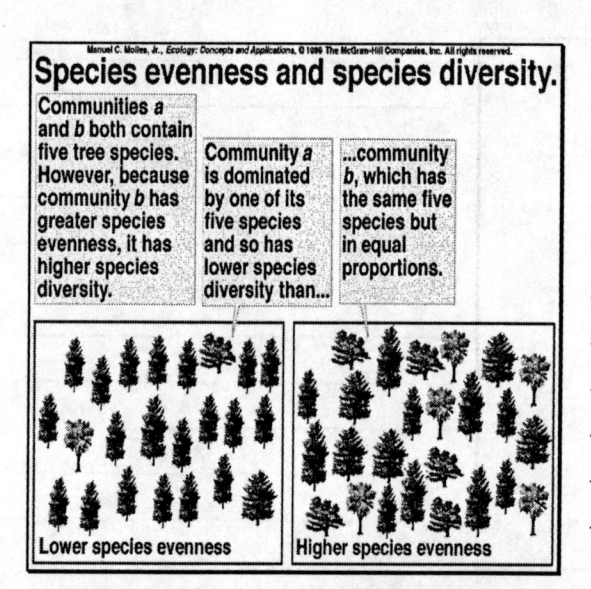

Species evenness and species diversity.

Communities *a* and *b* both contain five tree species. However, because community *b* has greater species evenness, it has higher species diversity.

Community *a* is dominated by one of its five species and so has lower species diversity than...

...community *b*, which has the same five species but in equal proportions.

Lower species evenness

Higher species evenness

Figure 13.6

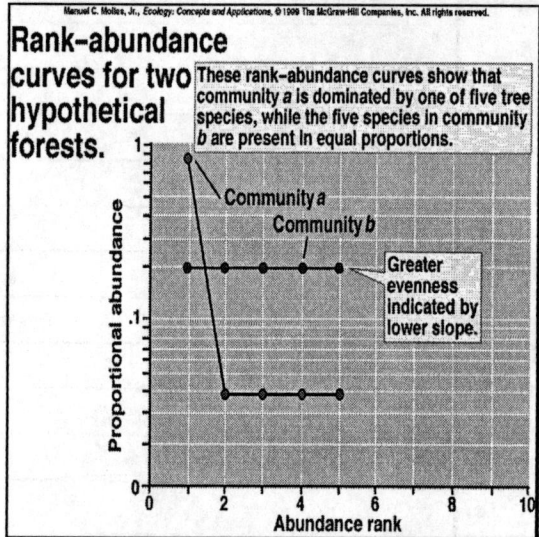

Rank-abundance curves for two hypothetical forests.

These rank-abundance curves show that community *a* is dominated by one of five tree species, while the five species in community *b* are present in equal proportions.

Community *a*
Community *b*
Greater evenness indicated by lower slope.

Figure 13.7

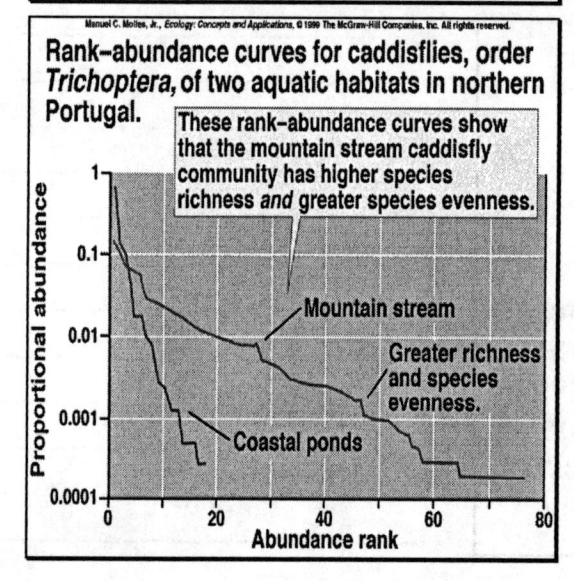

Rank-abundance curves for caddisflies, order *Trichoptera*, of two aquatic habitats in northern Portugal.

These rank-abundance curves show that the mountain stream caddisfly community has higher species richness *and* greater species evenness.

Mountain stream
Greater richness and species evenness.
Coastal ponds

Figure 13.8

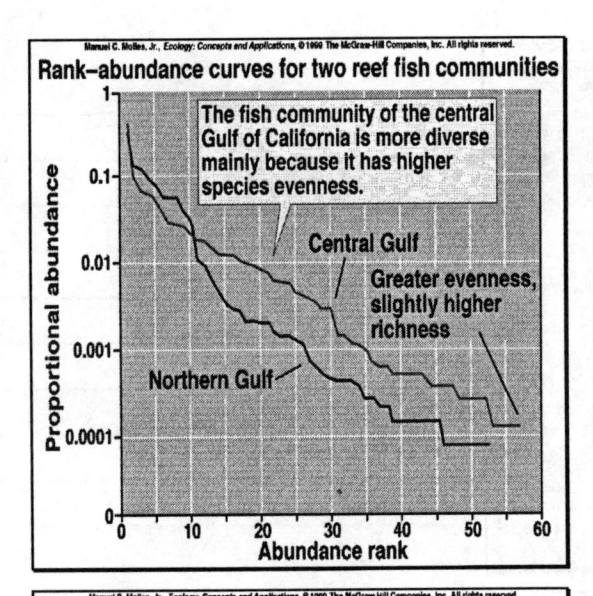

Rank–abundance curves for two reef fish communities

The fish community of the central Gulf of California is more diverse mainly because it has higher species evenness.

Central Gulf — Greater evenness, slightly higher richness

Northern Gulf

Proportional abundance vs. Abundance rank

Figure 13.9

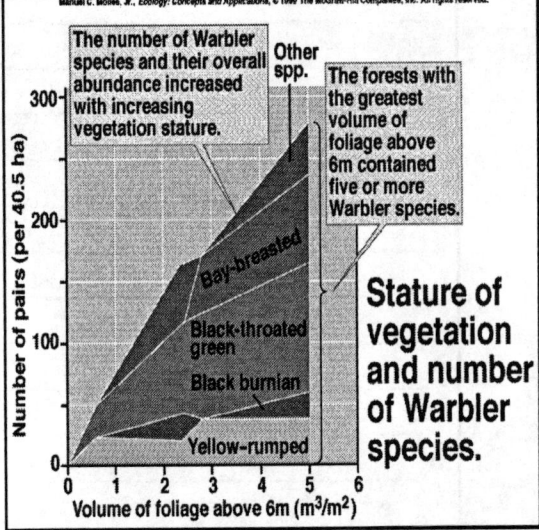

The number of Warbler species and their overall abundance increased with increasing vegetation stature.

The forests with the greatest volume of foliage above 6m contained five or more Warbler species.

Other spp.
Bay-breasted
Black-throated green
Black burnian
Yellow-rumped

Number of pairs (per 40.5 ha) vs. Volume of foliage above 6m (m^3/m^2)

Stature of vegetation and number of Warbler species.

Figure 13.10

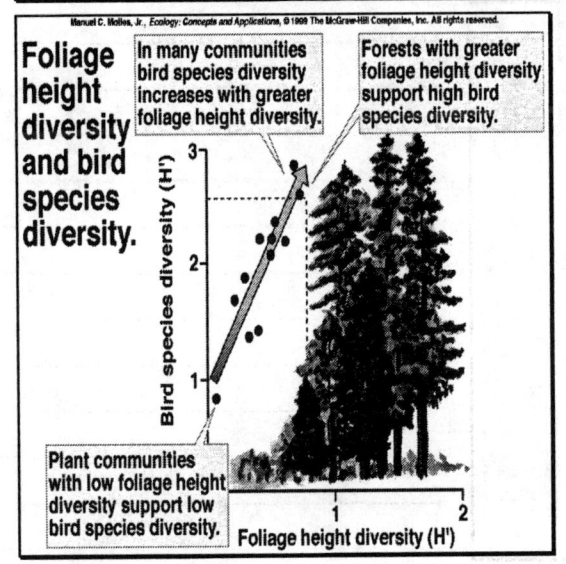

Foliage height diversity and bird species diversity.

In many communities bird species diversity increases with greater foliage height diversity.

Forests with greater foliage height diversity support high bird species diversity.

Plant communities with low foliage height diversity support low bird species diversity.

Bird species diversity (H') vs. Foliage height diversity (H')

Figure 13.11

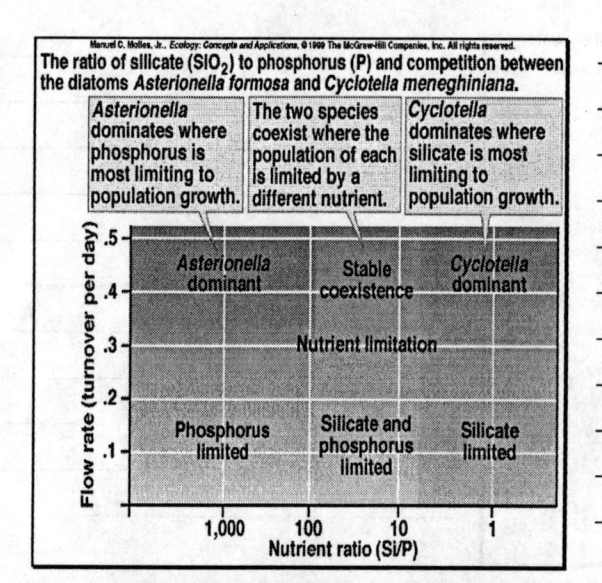

The ratio of silicate (SiO$_2$) to phosphorus (P) and competition between the diatoms *Asterionella formosa* and *Cyclotella meneghiniana*.

- *Asterionella* dominates where phosphorus is most limiting to population growth.
- The two species coexist where the population of each is limited by a different nutrient.
- *Cyclotella* dominates where silicate is most limiting to population growth.

Flow rate (turnover per day) vs. Nutrient ratio (Si/P)
- *Asterionella* dominant | Stable coexistence | *Cyclotella* dominant
- Nutrient limitation
- Phosphorus limited | Silicate and phosphorus limited | Silicate limited

Figure 13.12

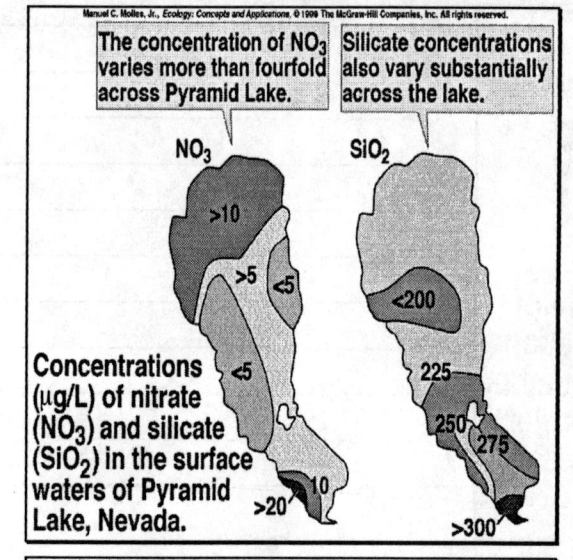

The concentration of NO$_3$ varies more than fourfold across Pyramid Lake. **Silicate concentrations also vary substantially across the lake.**

Concentrations (µg/L) of nitrate (NO$_3$) and silicate (SiO$_2$) in the surface waters of Pyramid Lake, Nevada.

NO$_3$: >10, >5, <5, <5, >20, 10
SiO$_2$: <200, 225, 250, 275, >300

Figure 13.13

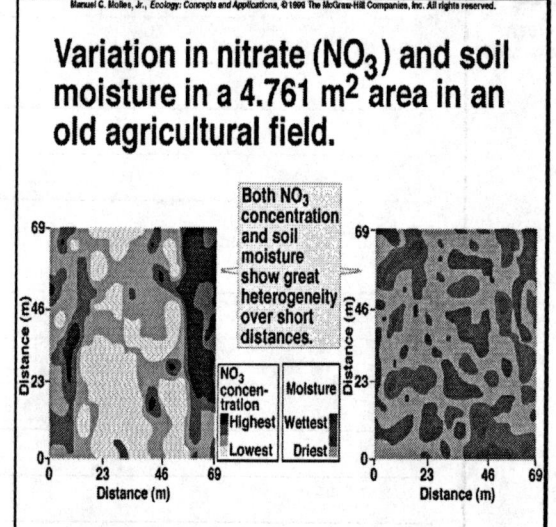

Variation in nitrate (NO$_3$) and soil moisture in a 4.761 m^2 area in an old agricultural field.

Both NO$_3$ concentration and soil moisture show great heterogeneity over short distances.

NO$_3$ concentration: Highest / Lowest
Moisture: Wettest / Driest

Figure 13.14

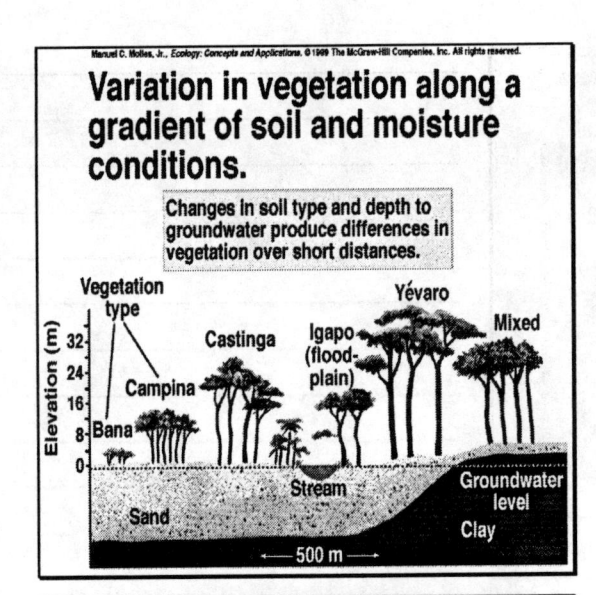

Variation in vegetation along a gradient of soil and moisture conditions.

Changes in soil type and depth to groundwater produce differences in vegetation over short distances.

Figure 13.15

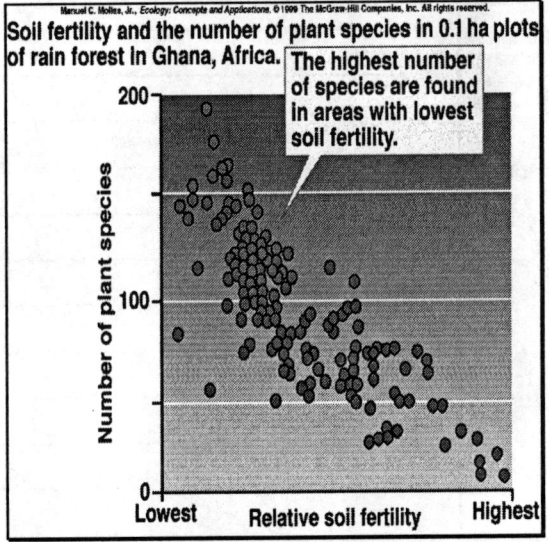

Soil fertility and the number of plant species in 0.1 ha plots of rain forest in Ghana, Africa.

The highest number of species are found in areas with lowest soil fertility.

Figure 13.16

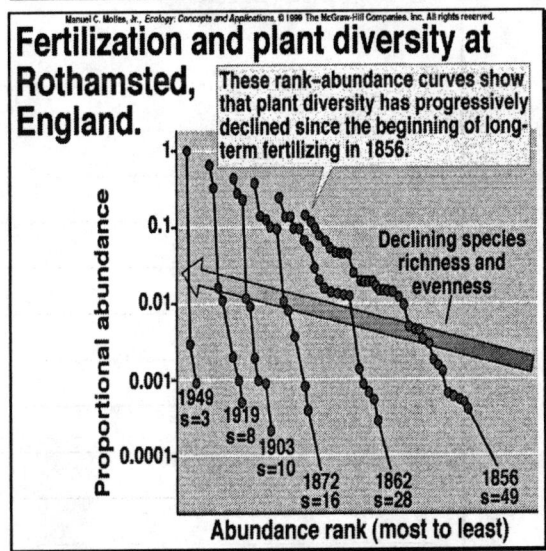

Fertilization and plant diversity at Rothamsted, England.

These rank–abundance curves show that plant diversity has progressively declined since the beginning of long-term fertilizing in 1856.

Declining species richness and evenness

1949 s=3
1919 s=8
1903 s=10
1872 s=16
1862 s=28
1856 s=49

Figure 13.17

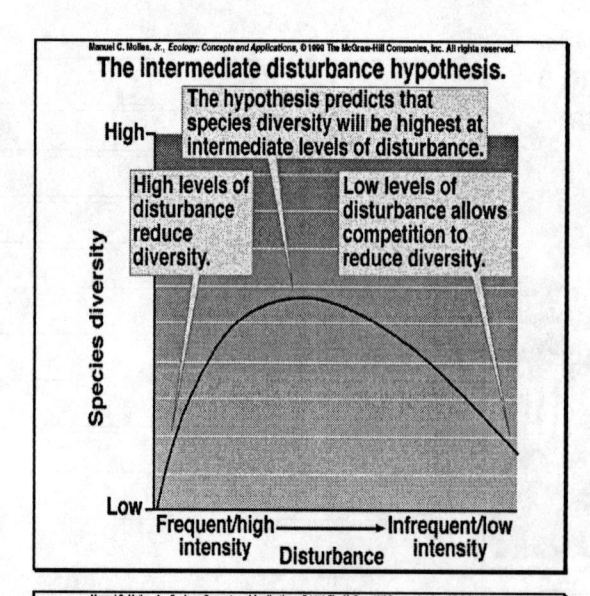

The intermediate disturbance hypothesis.

The hypothesis predicts that species diversity will be highest at intermediate levels of disturbance.

High levels of disturbance reduce diversity.

Low levels of disturbance allows competition to reduce diversity.

Species diversity — High / Low
Disturbance — Frequent/high intensity → Infrequent/low intensity

Figure 13.18

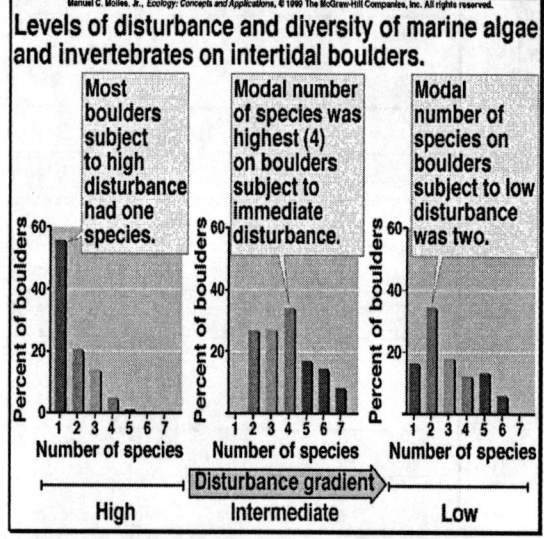

Levels of disturbance and diversity of marine algae and invertebrates on intertidal boulders.

Most boulders subject to high disturbance had one species.

Modal number of species was highest (4) on boulders subject to immediate disturbance.

Modal number of species on boulders subject to low disturbance was two.

Disturbance gradient: High — Intermediate — Low

Figure 13.19

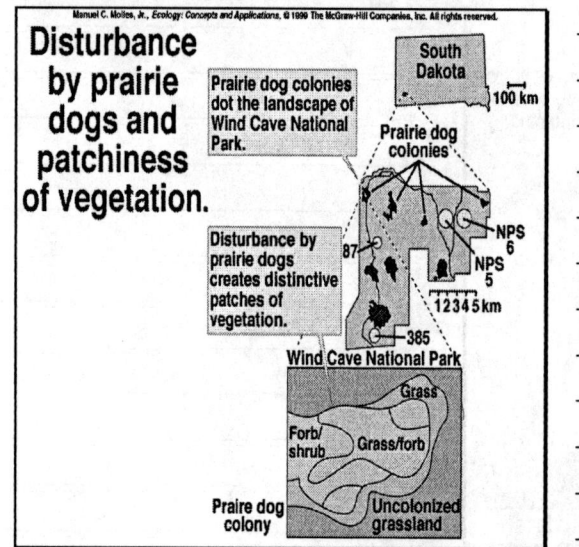

Disturbance by prairie dogs and patchiness of vegetation.

Prairie dog colonies dot the landscape of Wind Cave National Park.

Disturbance by prairie dogs creates distinctive patches of vegetation.

South Dakota

Wind Cave National Park

Forb/shrub, Grass, Grass/forb, Prairie dog colony, Uncolonized grassland

110

Figure 13.20

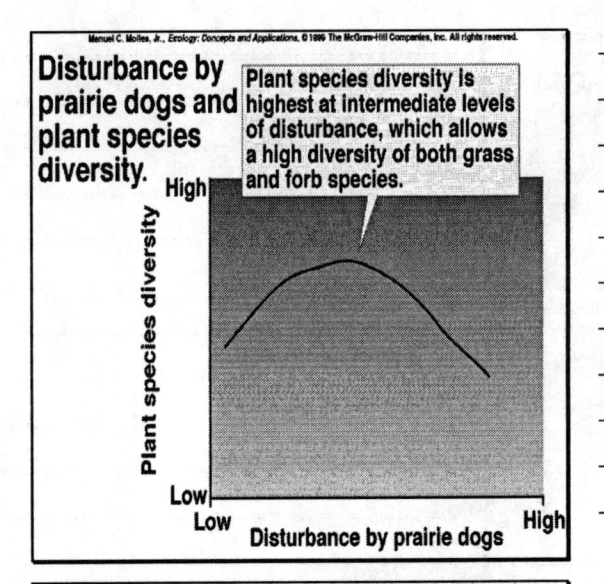

Disturbance by prairie dogs and plant species diversity.

Plant species diversity is highest at intermediate levels of disturbance, which allows a high diversity of both grass and forb species.

Figure 13.21 (I)

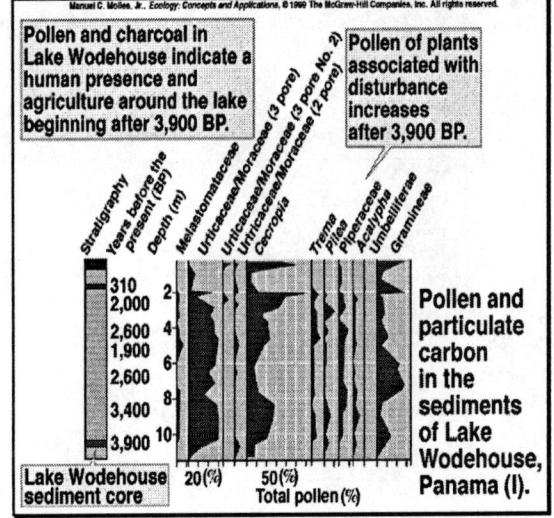

Pollen and charcoal in Lake Wodehouse indicate a human presence and agriculture around the lake beginning after 3,900 BP.

Pollen of plants associated with disturbance increases after 3,900 BP.

Pollen and particulate carbon in the sediments of Lake Wodehouse, Panama (I).

Figure 13.19 (II)

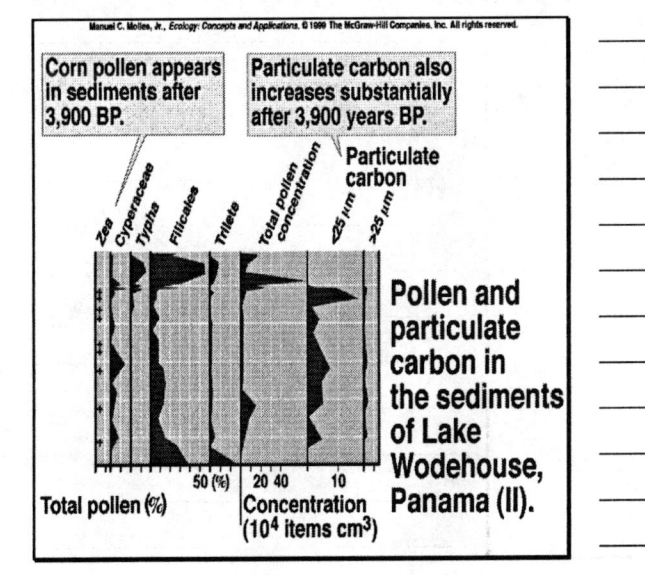

Corn pollen appears in sediments after 3,900 BP.

Particulate carbon also increases substantially after 3,900 years BP.

Pollen and particulate carbon in the sediments of Lake Wodehouse, Panama (II).

111

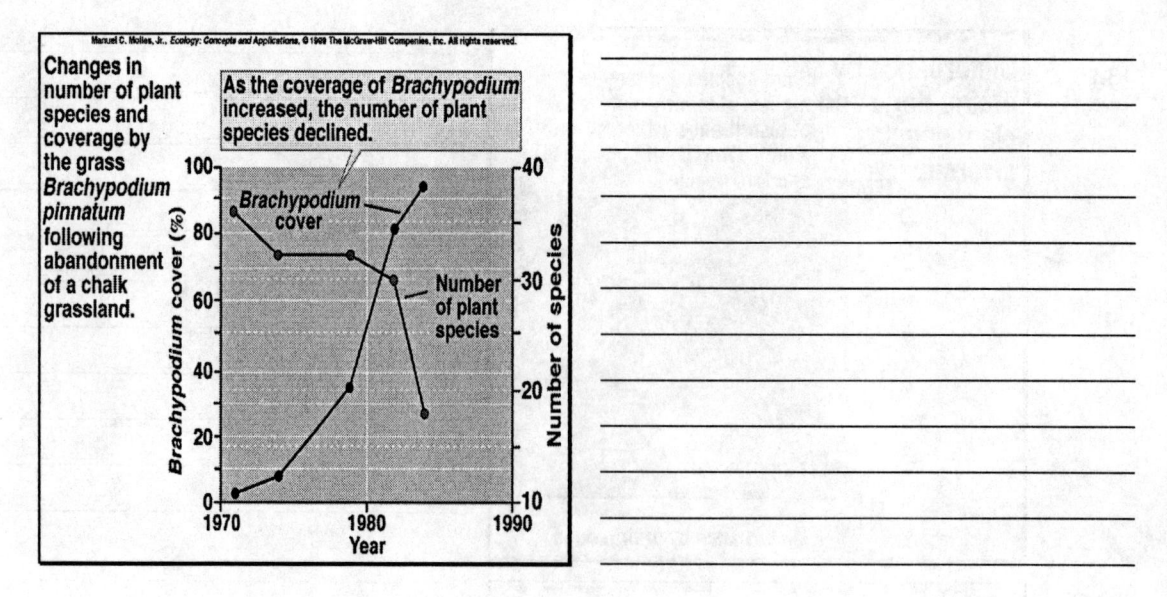

Figure 13.22

Changes in number of plant species and coverage by the grass *Brachypodium pinnatum* following abandonment of a chalk grassland.

As the coverage of *Brachypodium* increased, the number of plant species declined.

Chapter 14

Ecology
Concepts and Applications
First Edition

Manuel C. Molles, Jr.

(c) The McGraw-Hill Companies, Inc.

Figure 14.2

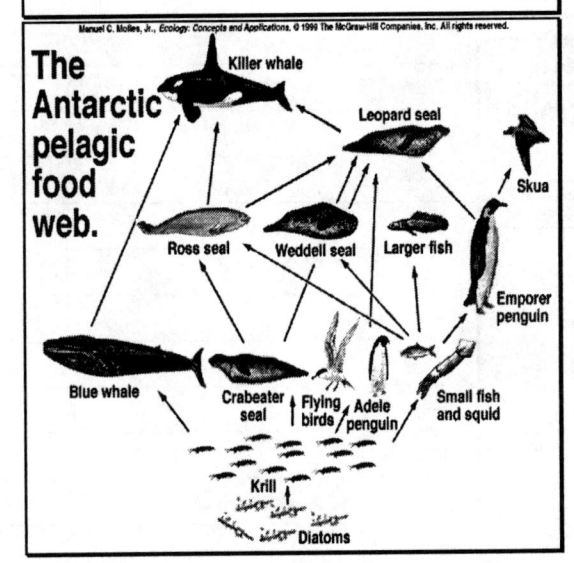

The Antarctic pelagic food web.

Figure 14.3

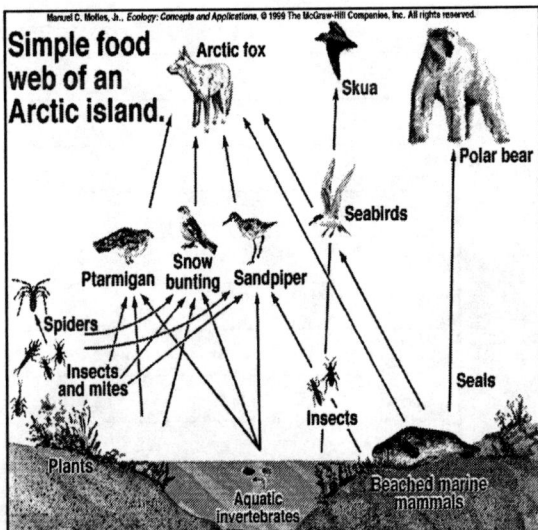

Simple food web of an Arctic island.

Figure 14.4

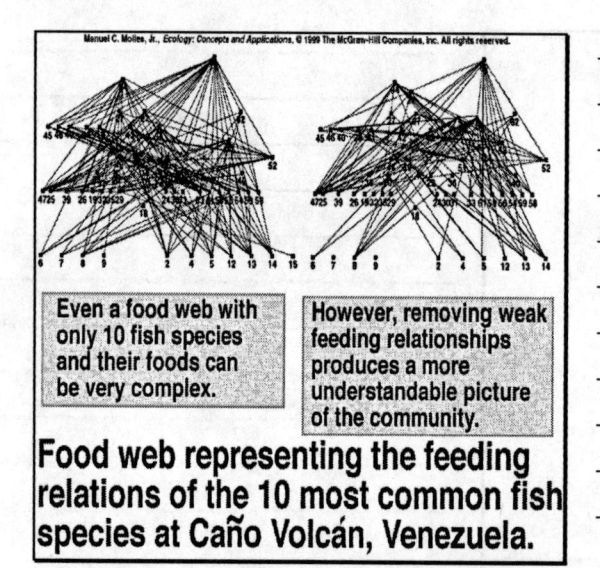

Food web representing the feeding relations of the 10 most common fish species at Caño Volcán, Venezuela.

Even a food web with only 10 fish species and their foods can be very complex.

However, removing weak feeding relationships produces a more understandable picture of the community.

Figure 14.5

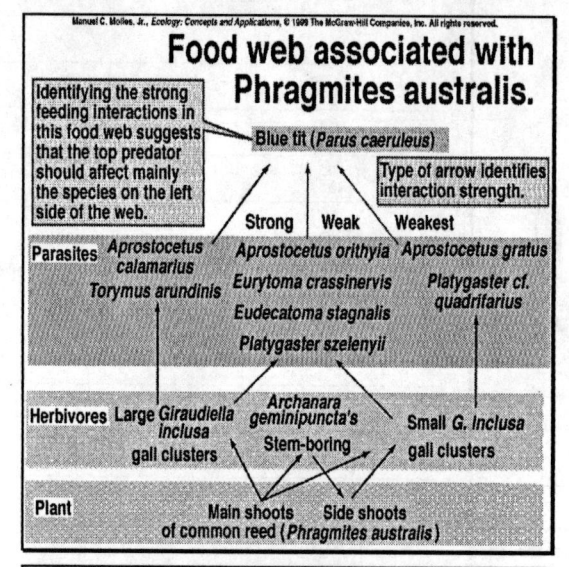

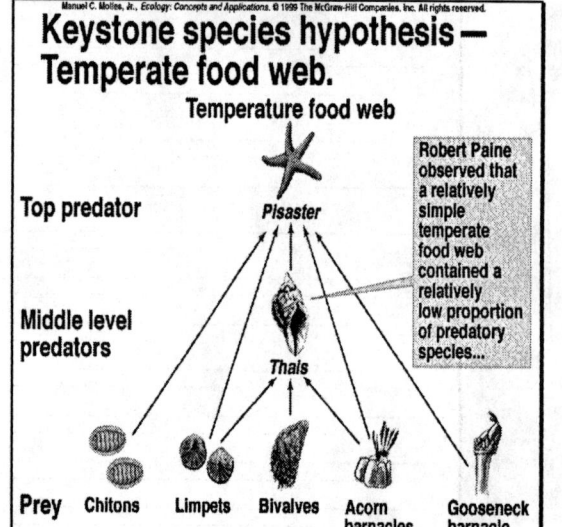

Food web associated with Phragmites australis.

Identifying the strong feeding interactions in this food web suggests that the top predator should affect mainly the species on the left side of the web.

Type of arrow identifies interaction strength.

Strong | Weak | Weakest

Blue tit (*Parus caeruleus*)

Parasites: *Aprostocetus calamarius*, *Torymus arundinis*, *Aprostocetus orithyia*, *Eurytoma crassinervis*, *Eudecatoma stagnalis*, *Platygaster szelenyii*, *Aprostocetus gratus*, *Platygaster* cf. *quadrifarius*

Herbivores: Large *Giraudiella inclusa* gall clusters, *Archanara geminipuncta's* Stem-boring, Small *G. inclusa* gall clusters

Plant: Main shoots, Side shoots of common reed (*Phragmites australis*)

Figure 14.6

Keystone species hypothesis — Temperate food web.

Temperature food web

Top predator: *Pisaster*

Middle level predators: *Thais*

Prey: Chitons, Limpets, Bivalves, Acorn barnacles, Gooseneck barnacle

Robert Paine observed that a relatively simple temperate food web contained a relatively low proportion of predatory species...

Figure 14.6

Keystone species hypothesis — subtropical food web.

...while a diverse subtropical food web contained a higher proportion of predatory species.

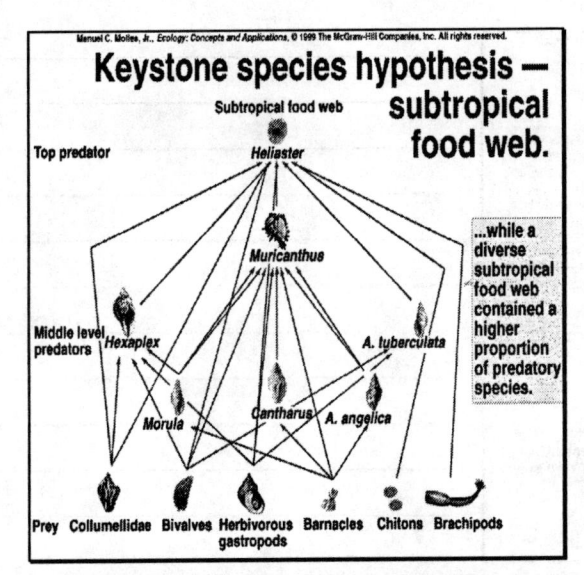

Figure 14.6

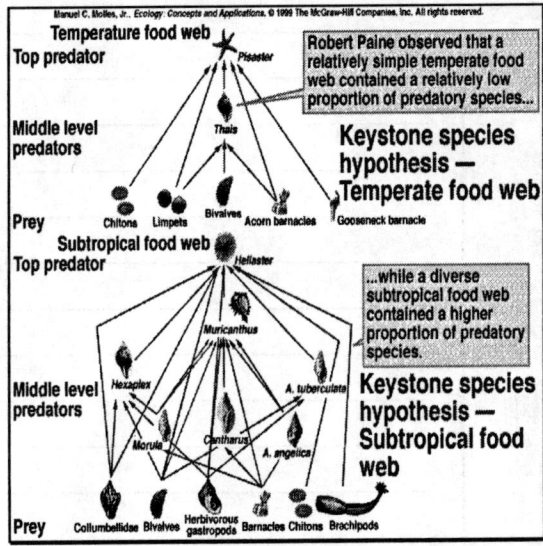

Temperate food web

Robert Paine observed that a relatively simple temperate food web contained a relatively low proportion of predatory species...

Keystone species hypothesis — Temperate food web

Subtropical food web

...while a diverse subtropical food web contained a higher proportion of predatory species.

Keystone species hypothesis — Subtropical food web

Figure 14.7

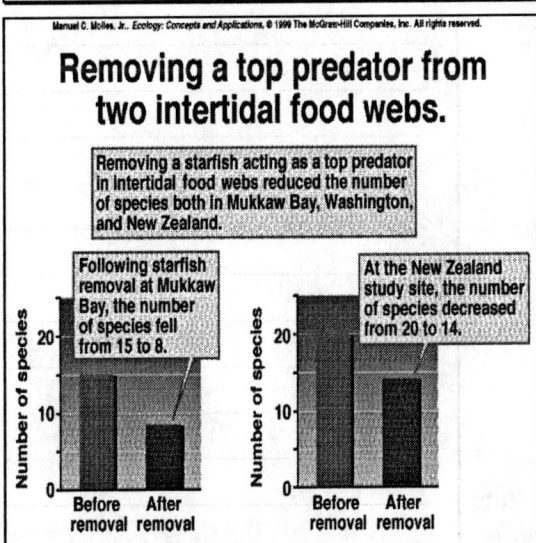

Removing a top predator from two intertidal food webs.

Removing a starfish acting as a top predator in intertidal food webs reduced the number of species both in Mukkaw Bay, Washington, and New Zealand.

Following starfish removal at Mukkaw Bay, the number of species fell from 15 to 8.

At the New Zealand study site, the number of species decreased from 20 to 14.

Figure 14.8: Effect of *Littorina littorea* on algal communities in tide pools.

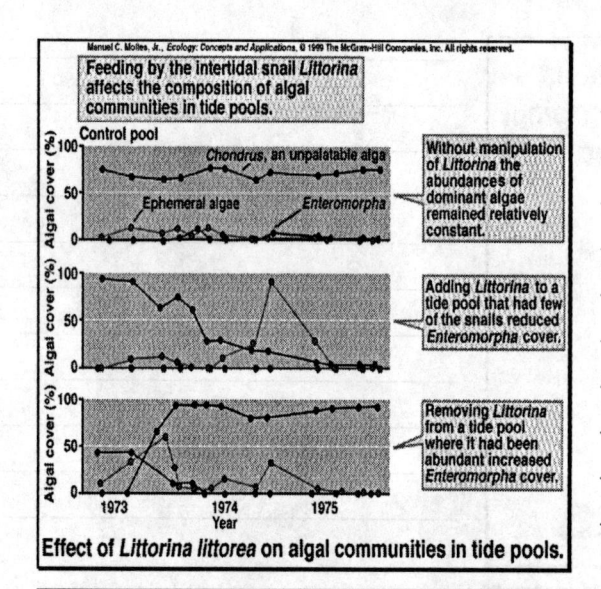

Figure 14.9: Effect of *Littorina littorea* on algal species richness in tide pools and emergent habitats.

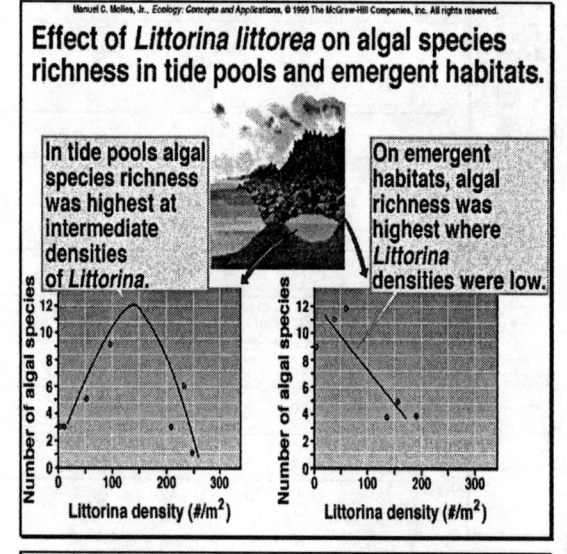

Figure 14.11: Food web associated with algal turf during the summer in the Eel River, California.

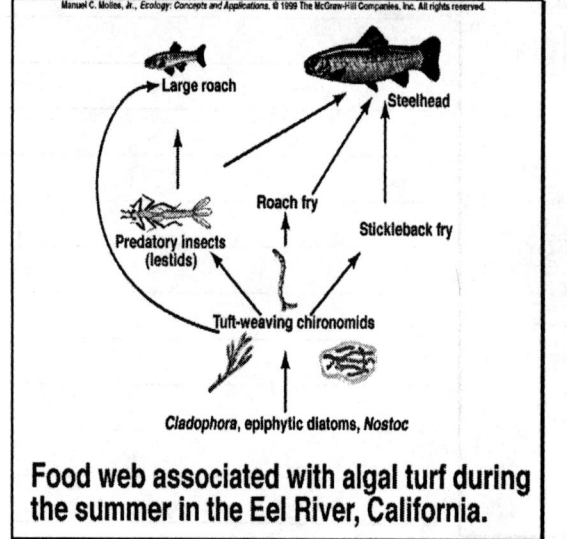

Figure 14.12

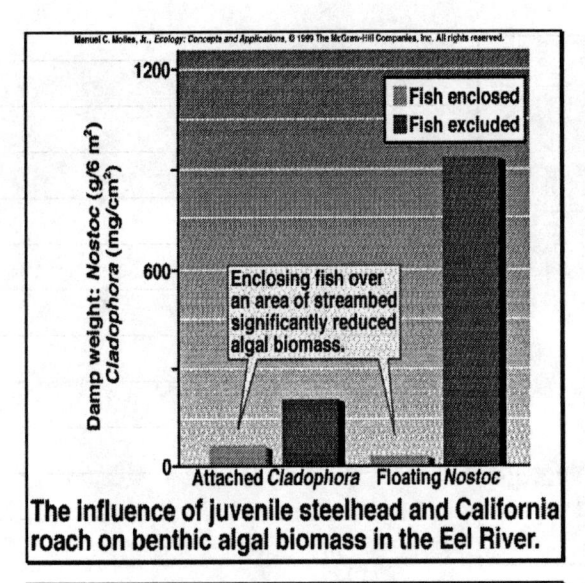

The influence of juvenile steelhead and California roach on benthic algal biomass in the Eel River.

Figure 14.13

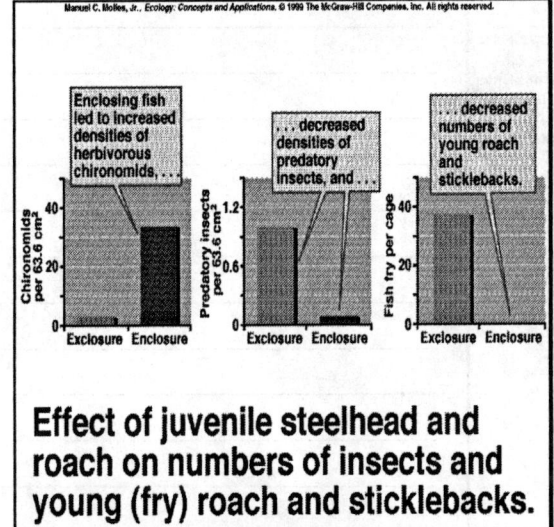

Effect of juvenile steelhead and roach on numbers of insects and young (fry) roach and sticklebacks.

Figure 14.14

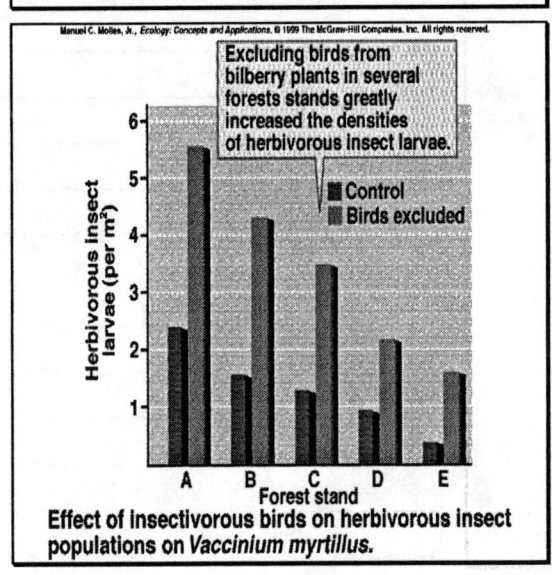

Effect of insectivorous birds on herbivorous insect populations on *Vaccinium myrtillus*.

Figure 14.15

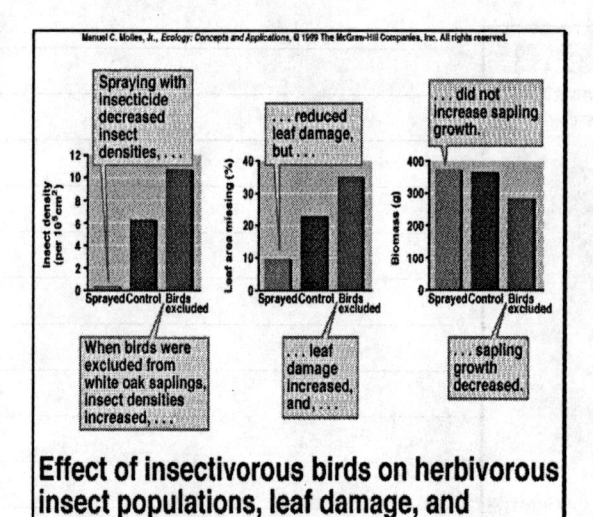

Effect of insectivorous birds on herbivorous insect populations, leaf damage, and sapling growth in white oaks.

Figure 14.16

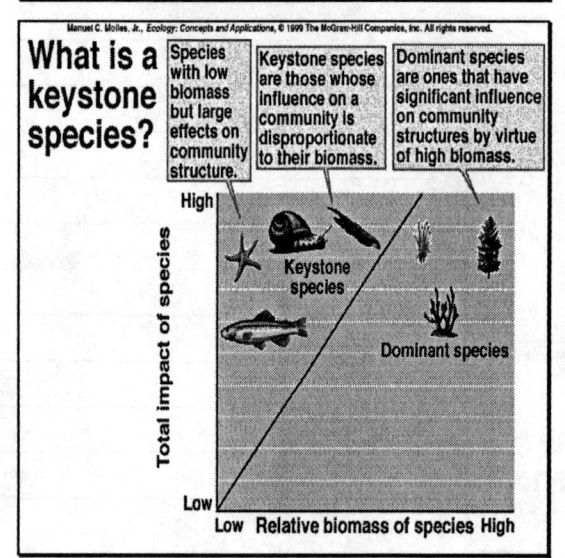

What is a keystone species?

Figure 14.17

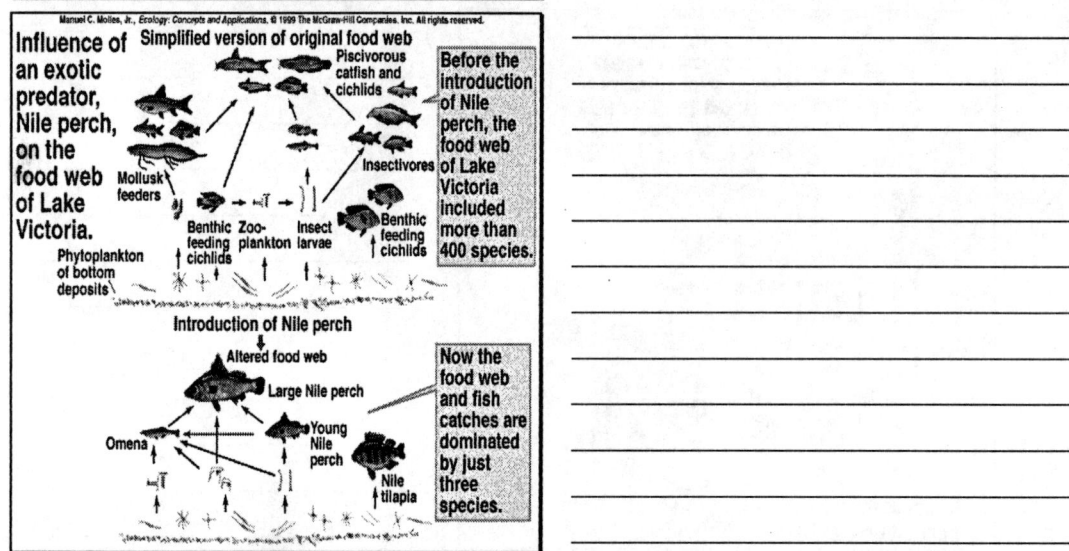

Influence of an exotic predator, Nile perch, on the food web of Lake Victoria.

Figure 14.18

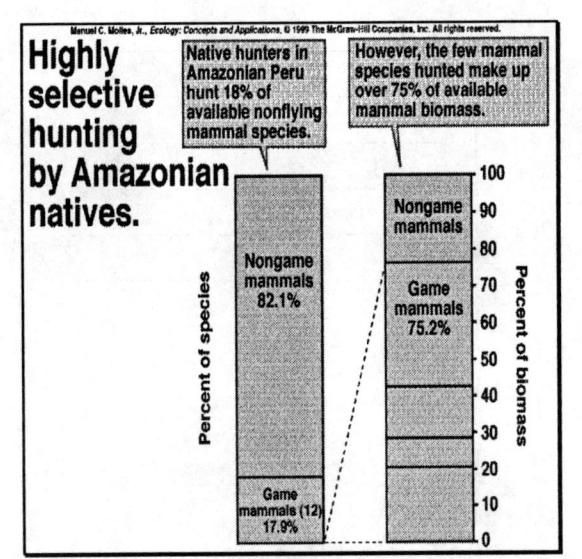

Figure 14.20

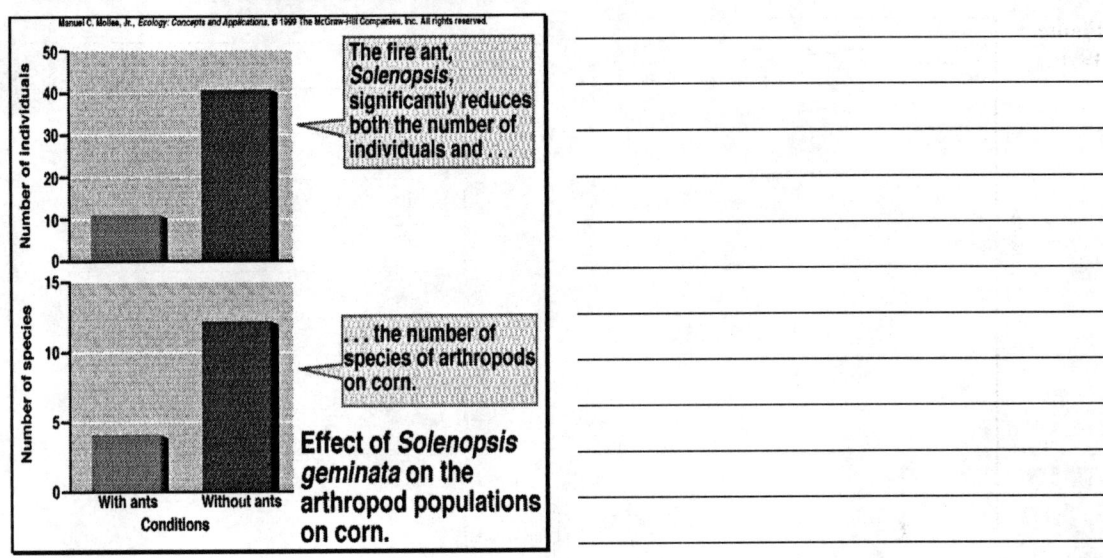

Chapter 15

Ecology

Concepts and Applications
First Edition

Manuel C. Molles, Jr.

(c) The McGraw-Hill Companies, Inc.

Figure 15.2

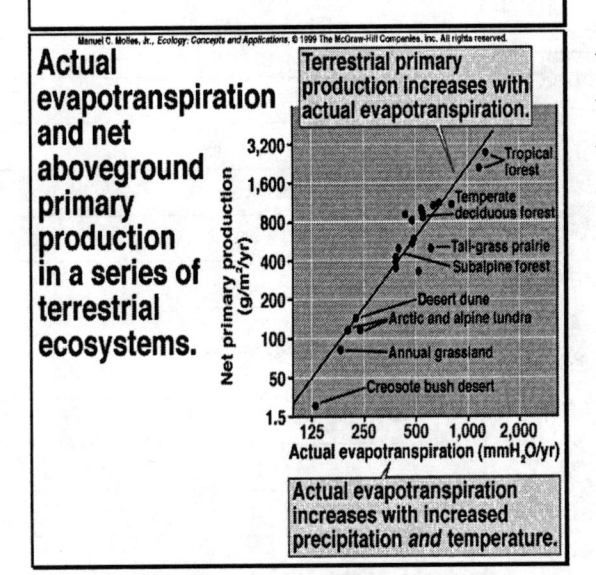

Actual evapotranspiration and net aboveground primary production in a series of terrestrial ecosystems.

Terrestrial primary production increases with actual evapotranspiration.

Actual evapotranspiration increases with increased precipitation *and* temperature.

Figure 15.3

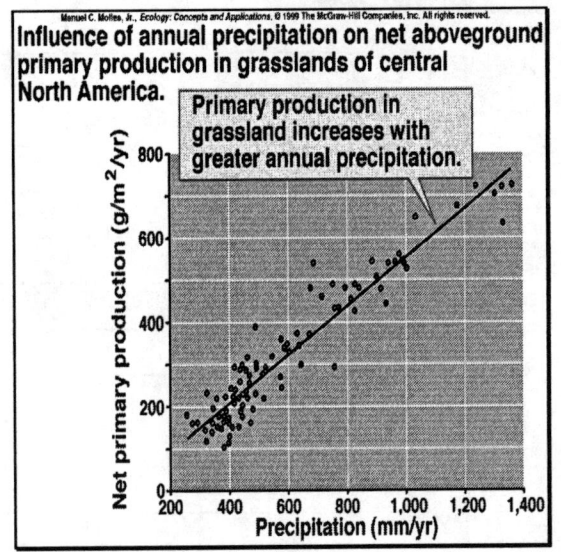

Influence of annual precipitation on net aboveground primary production in grasslands of central North America.

Primary production in grassland increases with greater annual precipitation.

Figure 15.4 — Adding nitrogen, phosphorus, and potassium to net aboveground primary production in Arctic tundra.

Adding fertilizers nearly doubled primary production in these tundra study plots.

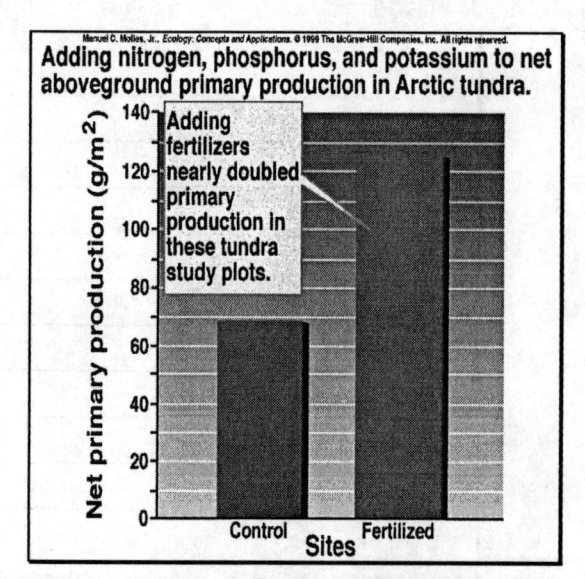

Figure 15.5 — Adding phosphorus (P) and/or nitrogen (N) to aboveground primary production in alpine tundra.

Adding nutrients increased primary production in both dry and wet meadows.

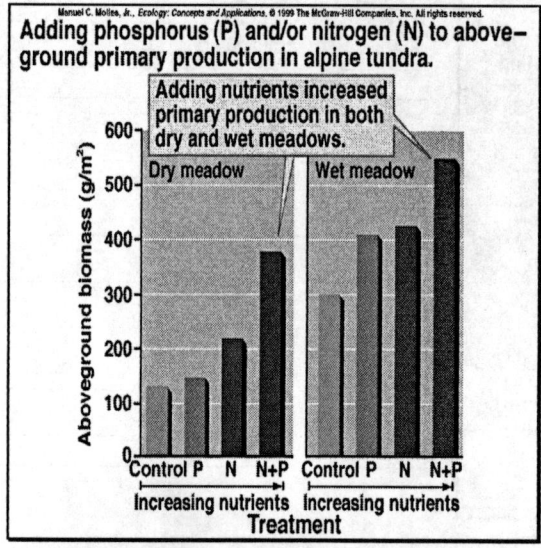

Figure 15.6 — Phosphorus concentration and algal biomass in north temperate lakes.

Higher phosphorus concentrations are associated with greater algal biomass.

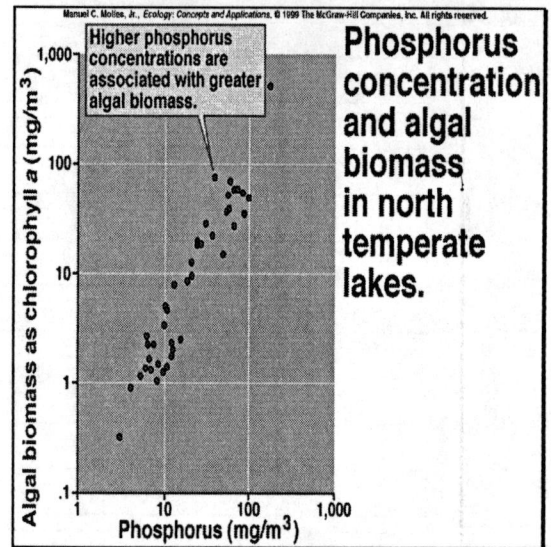

Figure 15.7

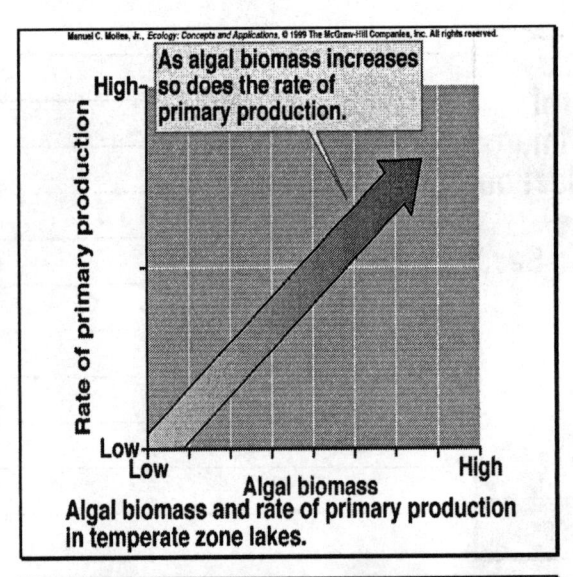

Algal biomass and rate of primary production in temperate zone lakes.

Figure 15.8

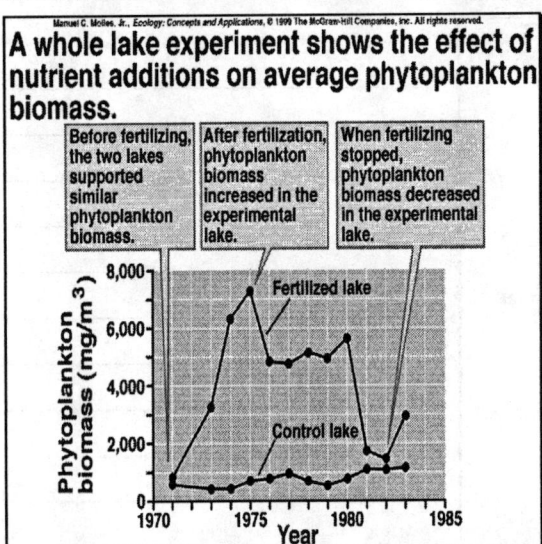

A whole lake experiment shows the effect of nutrient additions on average phytoplankton biomass.

Figure 15.9

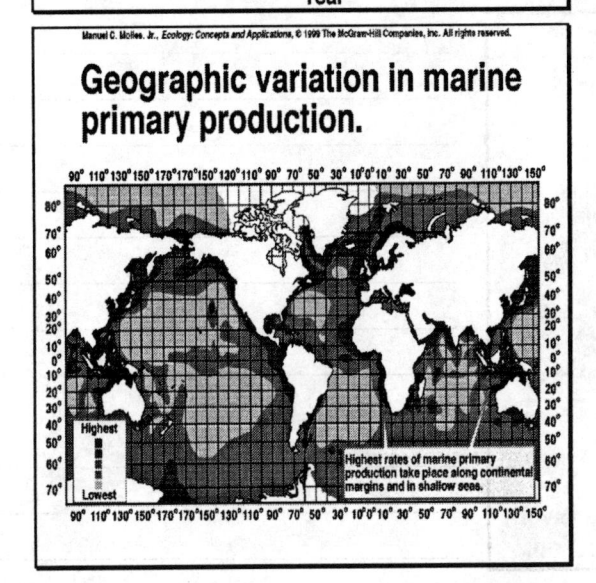

Geographic variation in marine primary production.

Figure 15.10

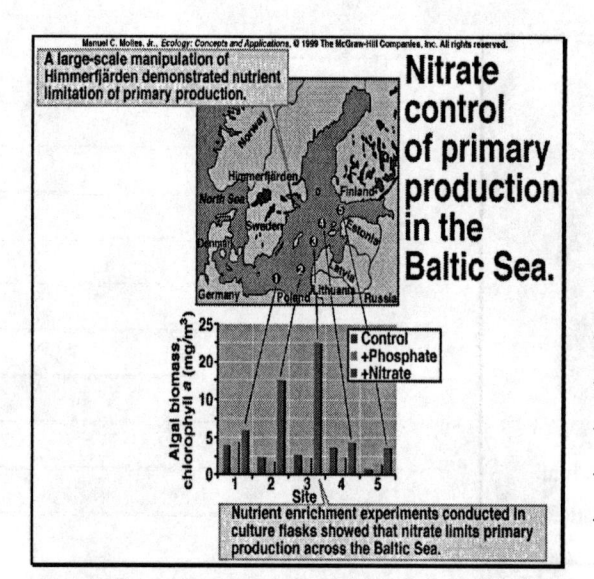

Nitrate control of primary production in the Baltic Sea.

A large-scale manipulation of Himmerfjärden demonstrated nutrient limitation of primary production.

Nutrient enrichment experiments conducted in culture flasks showed that nitrate limits primary production across the Baltic Sea.

Figure 15.11

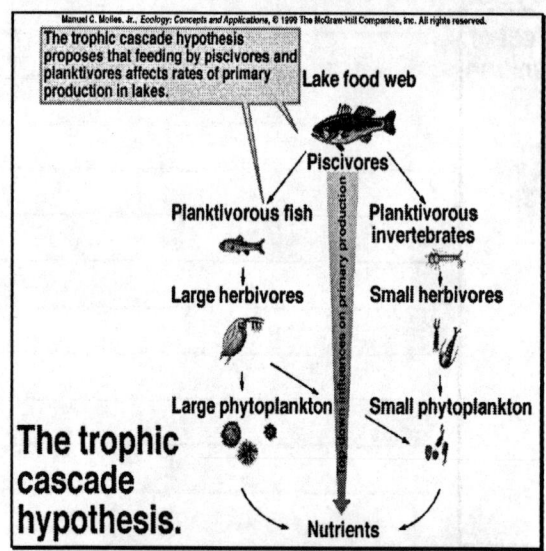

The trophic cascade hypothesis.

The trophic cascade hypothesis proposes that feeding by piscivores and planktivores affects rates of primary production in lakes.

Lake food web: Piscivores → Planktivorous fish / Planktivorous invertebrates → Large herbivores / Small herbivores → Large phytoplankton / Small phytoplankton → Nutrients

Figure 15.12

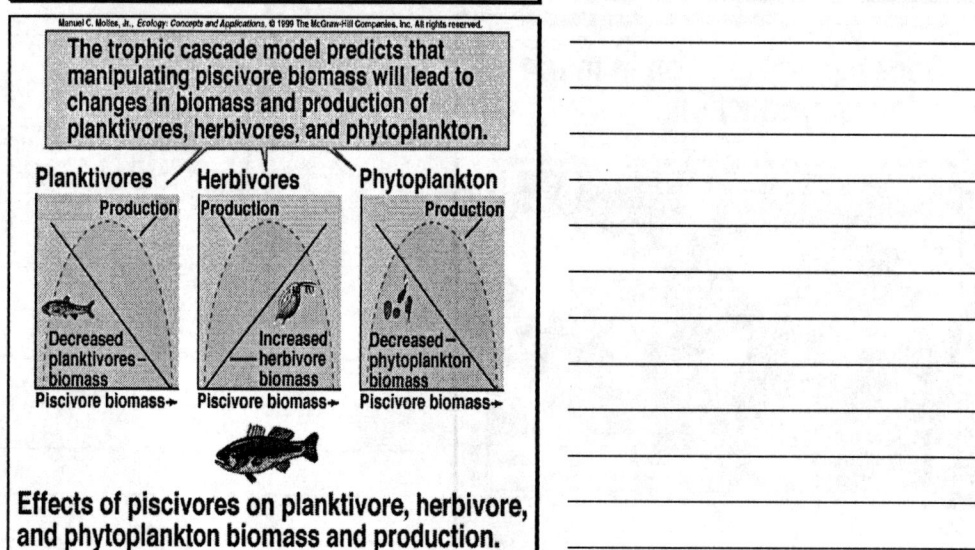

The trophic cascade model predicts that manipulating piscivore biomass will lead to changes in biomass and production of planktivores, herbivores, and phytoplankton.

Planktivores: Decreased planktivores biomass vs Piscivore biomass
Herbivores: Increased herbivore biomass vs Piscivore biomass
Phytoplankton: Decreased phytoplankton biomass vs Piscivore biomass

Effects of piscivores on planktivore, herbivore, and phytoplankton biomass and production.

Figure 15.13

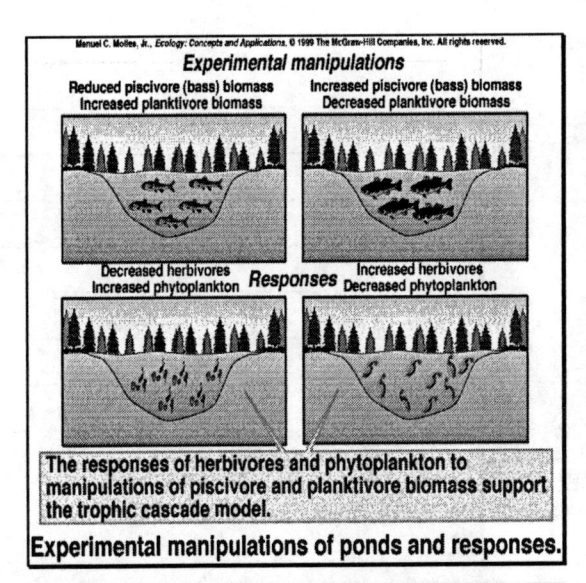

Experimental manipulations

Reduced piscivore (bass) biomass / Increased planktivore biomass

Increased piscivore (bass) biomass / Decreased planktivore biomass

Responses

Decreased herbivores / Increased phytoplankton

Increased herbivores / Decreased phytoplankton

The responses of herbivores and phytoplankton to manipulations of piscivore and planktivore biomass support the trophic cascade model.

Experimental manipulations of ponds and responses.

Figure 15.14

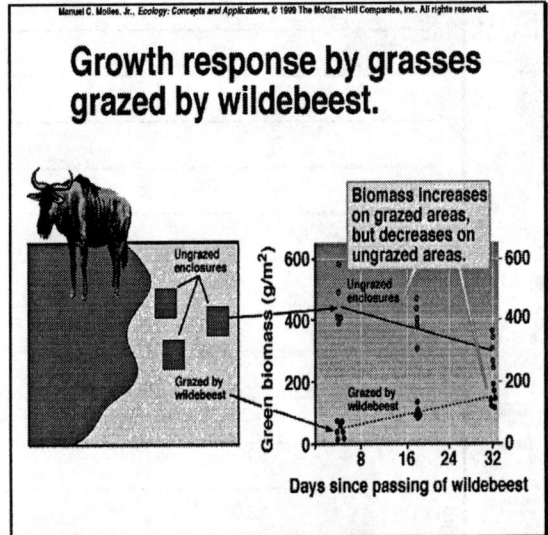

Growth response by grasses grazed by wildebeest.

Biomass increases on grazed areas, but decreases on ungrazed areas.

Ungrazed enclosures

Grazed by wildebeest

Green biomass (g/m^2)

Days since passing of wildebeest

Figure 15.15

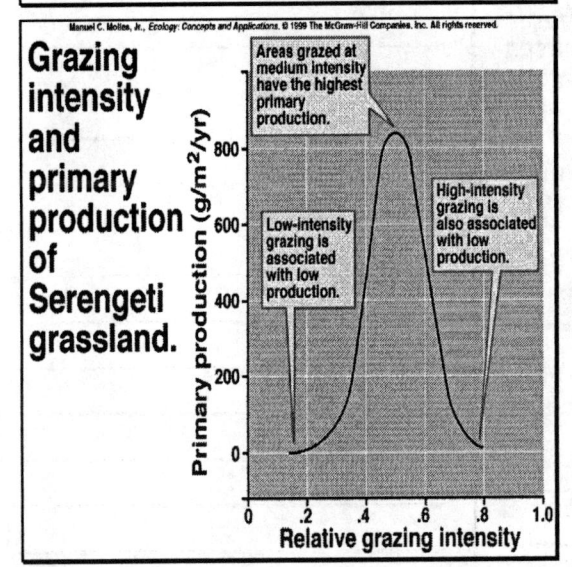

Grazing intensity and primary production of Serengeti grassland.

Areas grazed at medium intensity have the highest primary production.

Low-intensity grazing is associated with low production.

High-intensity grazing is also associated with low production.

Primary production (g/m^2/yr)

Relative grazing intensity

125

Figure 15.16

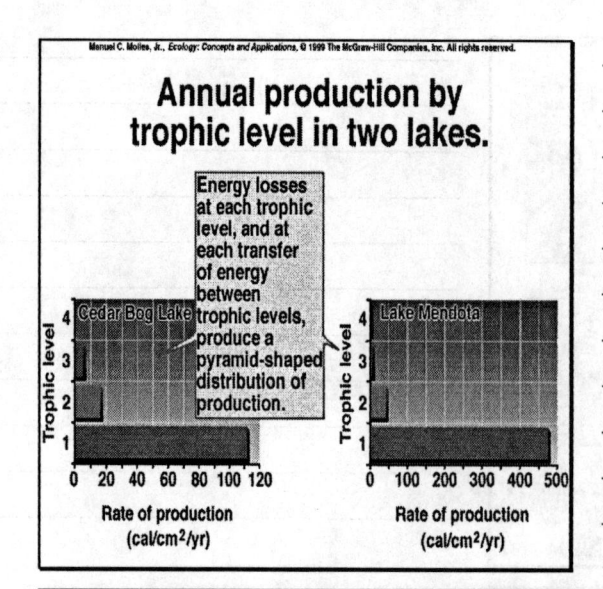

Annual production by trophic level in two lakes.

Energy losses at each trophic level, and at each transfer of energy between trophic levels, produce a pyramid-shaped distribution of production.

Figure 15.17

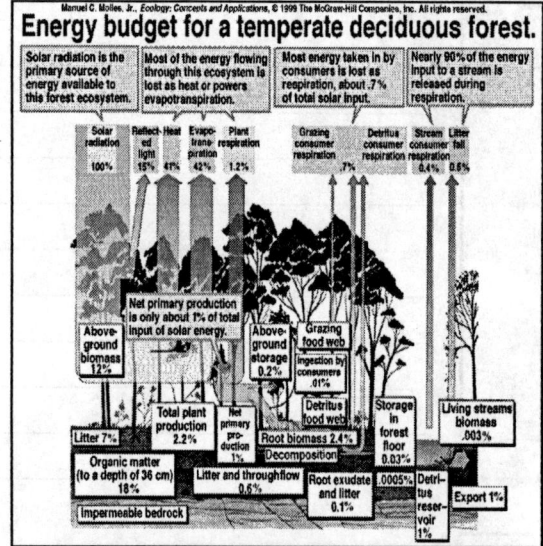

Energy budget for a temperate deciduous forest.

Figure 15.18

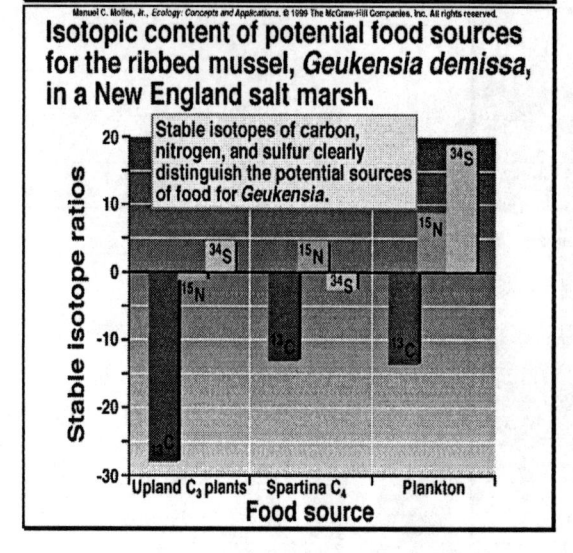

Isotopic content of potential food sources for the ribbed mussel, *Geukensia demissa*, in a New England salt marsh.

Stable isotopes of carbon, nitrogen, and sulfur clearly distinguish the potential sources of food for *Geukensia*.

Figure 15.19

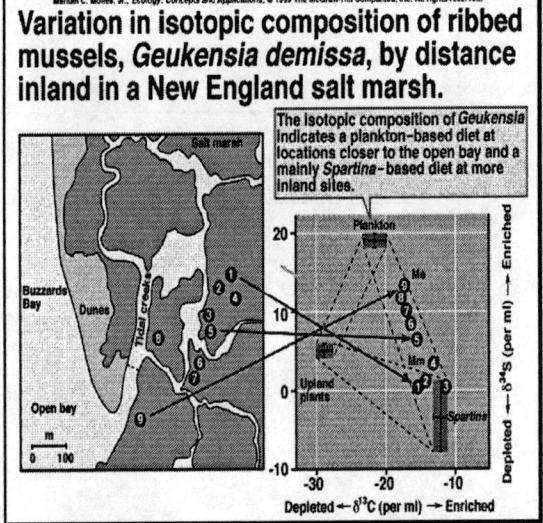

Variation in isotopic composition of ribbed mussels, *Geukensia demissa*, by distance inland in a New England salt marsh.

The isotopic composition of *Geukensia* indicates a plankton-based diet at locations closer to the open bay and a mainly *Spartina*-based diet at more inland sites.

Figure 15.20

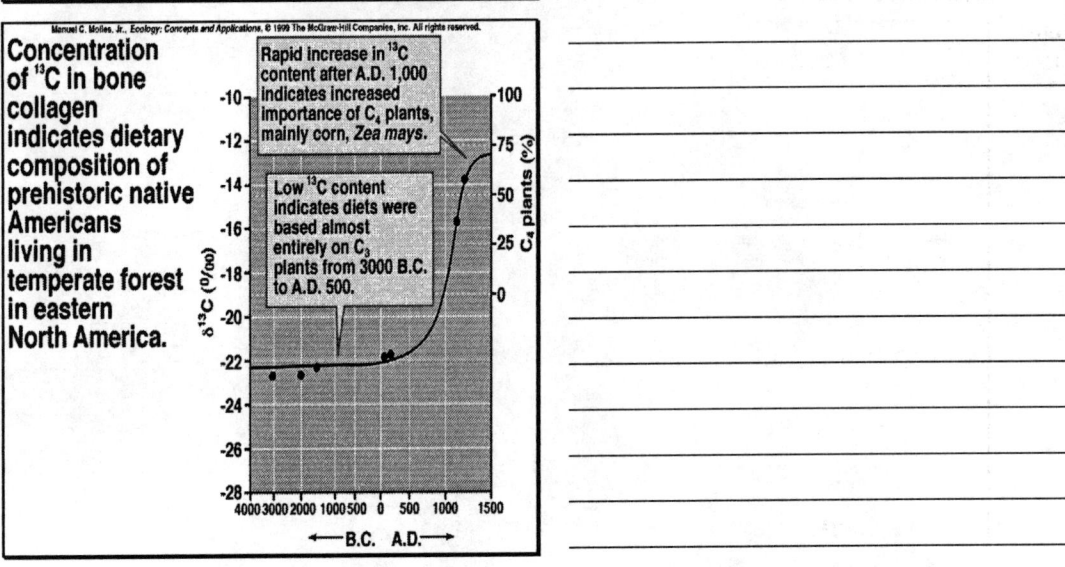

Concentration of ^{13}C in bone collagen indicates dietary composition of prehistoric native Americans living in temperate forest in eastern North America.

Rapid increase in ^{13}C content after A.D. 1,000 indicates increased importance of C_4 plants, mainly corn, *Zea mays*.

Low ^{13}C content indicates diets were based almost entirely on C_3 plants from 3000 B.C. to A.D. 500.

127

Chapter 16

Ecology
Concepts and Applications
First Edition

Manuel C. Molles, Jr.

(c) The McGraw-Hill Companies, Inc.

Figure 16.1

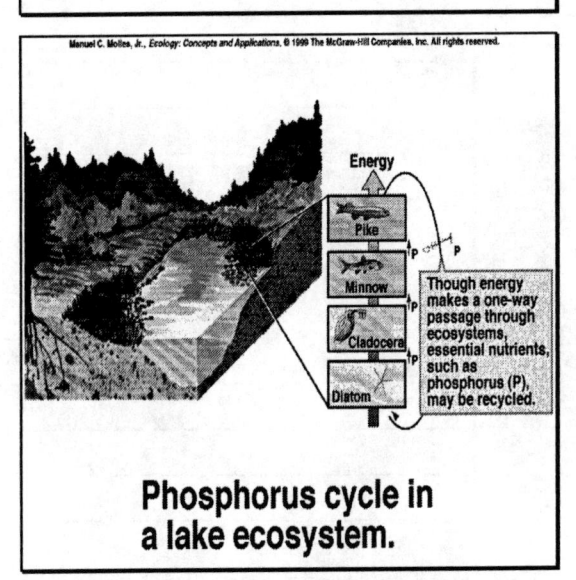

Phosphorus cycle in a lake ecosystem.

Though energy makes a one-way passage through ecosystems, essential nutrients, such as phosphorus (P), may be recycled.

Figure 16.2

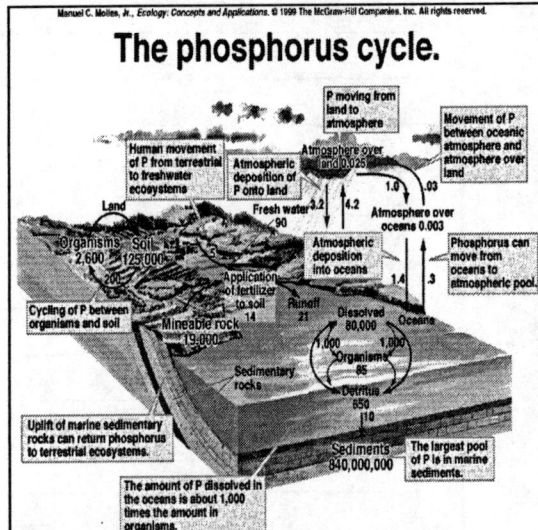

The phosphorus cycle.

Figure 16.3: The nitrogen cycle.

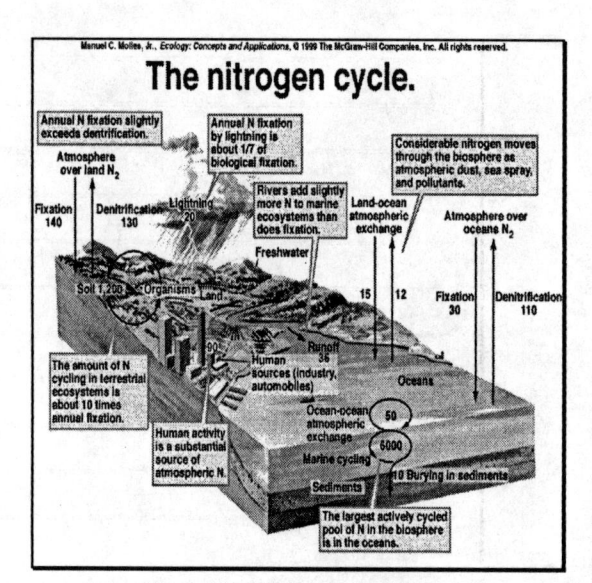

Figure 16.4: The carbon cycle.

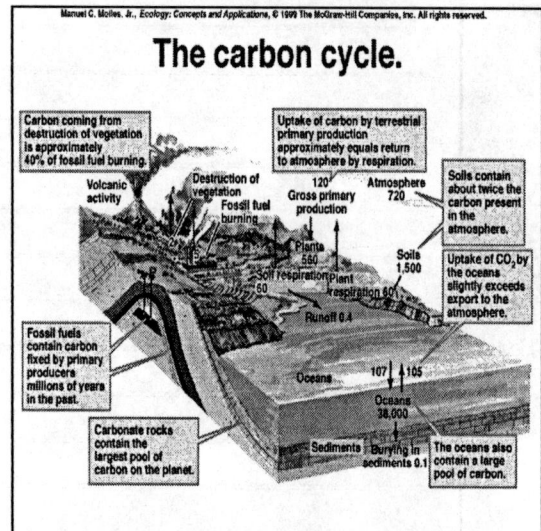

Figure 16.5: Decomposition of *Fraxinus angustifolia* leaves at wetter and drier sites.

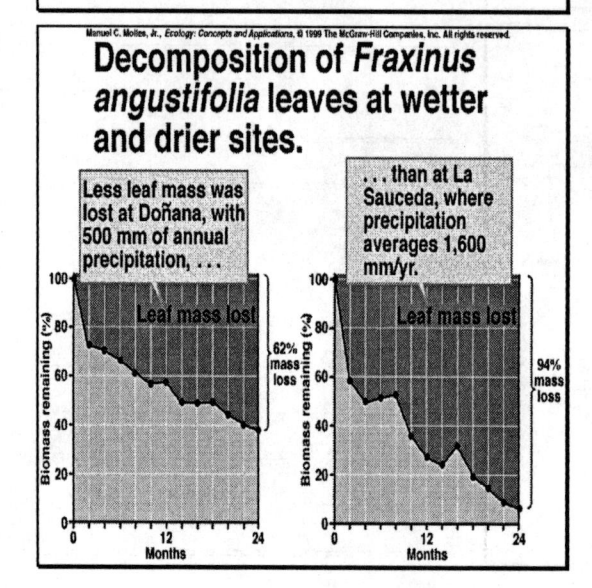

Figure 16.6

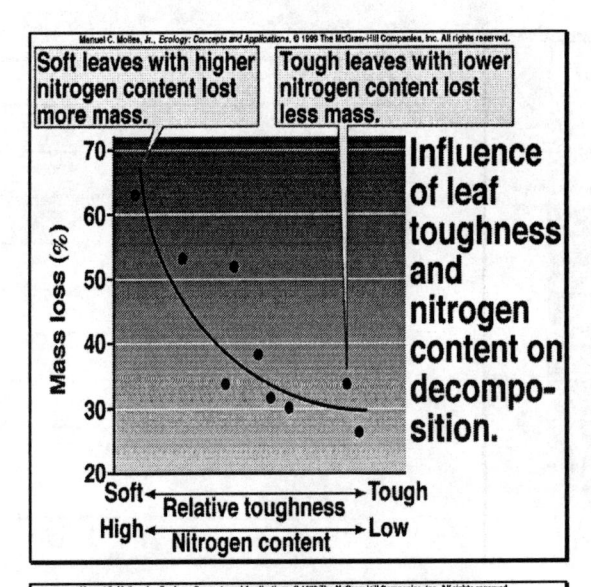

Soft leaves with higher nitrogen content lost more mass.

Tough leaves with lower nitrogen content lost less mass.

Influence of leaf toughness and nitrogen content on decomposition.

Mass loss (%)

Soft ← Relative toughness → Tough
High ← Nitrogen content → Low

Figure 16.7

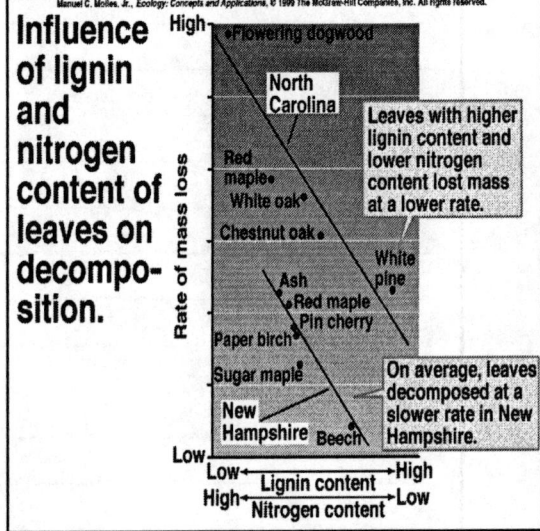

Influence of lignin and nitrogen content of leaves on decomposition.

Rate of mass loss

- Flowering dogwood
- North Carolina
- Red maple
- White oak
- Chestnut oak
- Ash
- Red maple
- Pin cherry
- White pine
- Paper birch
- Sugar maple
- New Hampshire
- Beech

Leaves with higher lignin content and lower nitrogen content lost mass at a lower rate.

On average, leaves decomposed at a slower rate in New Hampshire.

Low ← Lignin content → High
High ← Nitrogen content → Low

Figure 16.8

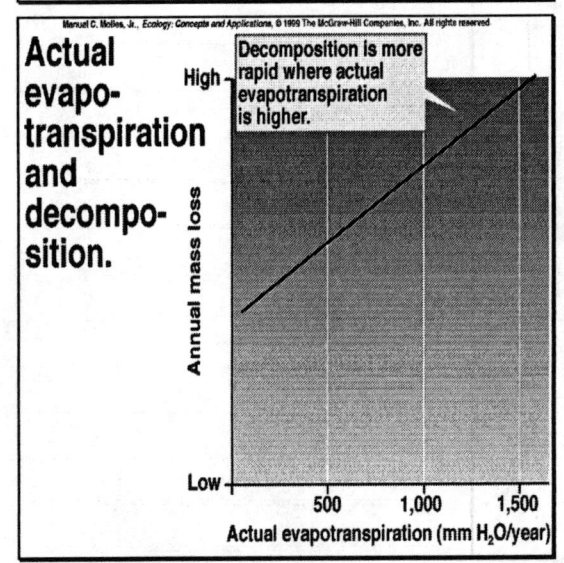

Actual evapotranspiration and decomposition.

Decomposition is more rapid where actual evapotranspiration is higher.

Annual mass loss

Actual evapotranspiration (mm H$_2$O/year)

131

Figure 16.9

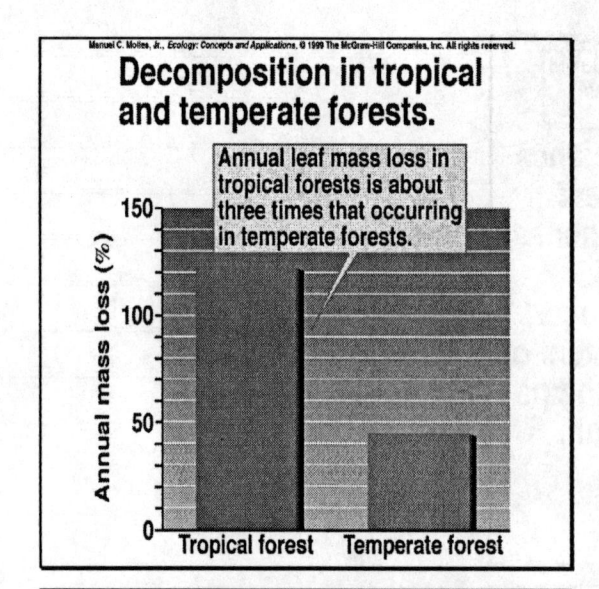

Decomposition in tropical and temperate forests.

Annual leaf mass loss in tropical forests is about three times that occurring in temperate forests.

Figure 16.10

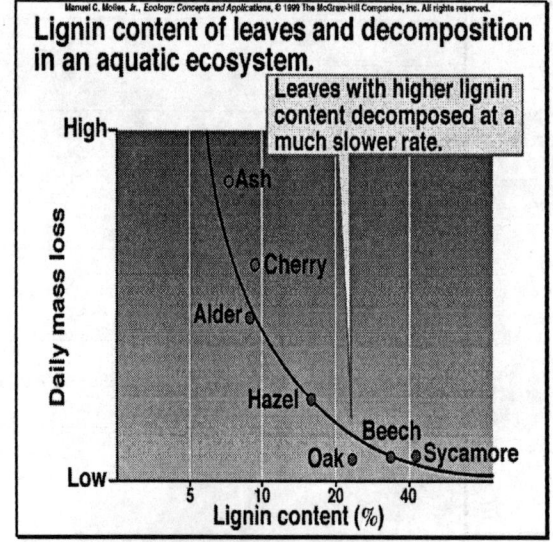

Lignin content of leaves and decomposition in an aquatic ecosystem.

Leaves with higher lignin content decomposed at a much slower rate.

Figure 16.11

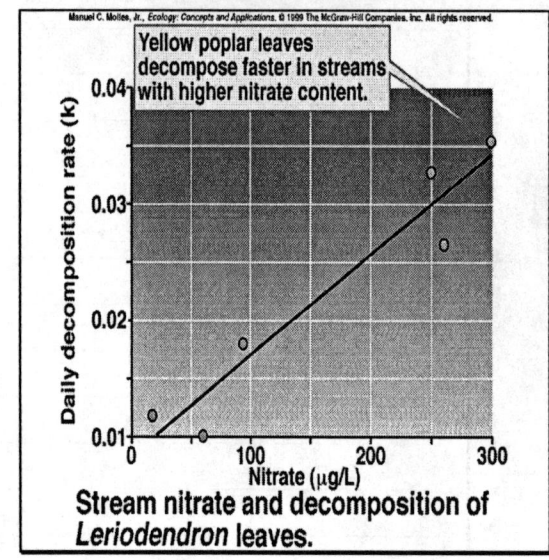

Yellow poplar leaves decompose faster in streams with higher nitrate content.

Stream nitrate and decomposition of *Leriodendron* leaves.

132

Figure 16.12

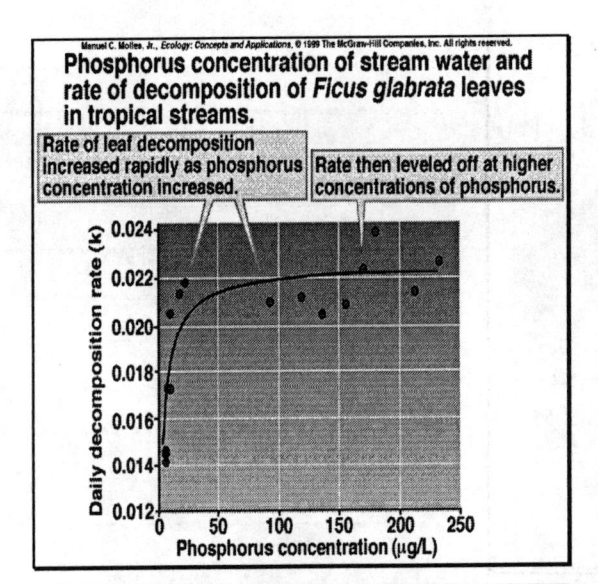

Phosphorus concentration of stream water and rate of decomposition of *Ficus glabrata* leaves in tropical streams.

Rate of leaf decomposition increased rapidly as phosphorus concentration increased.

Rate then leveled off at higher concentrations of phosphorus.

Figure 16.13

Nutrient spiraling in streams.

The transport of nutrients by streams is slowed by nutrient uptake by the benthic ecosystem.

Nutrient release from the benthos transport by water, and downstream uptake by the benthos produce nutrient spiraling.

Figure 16.14

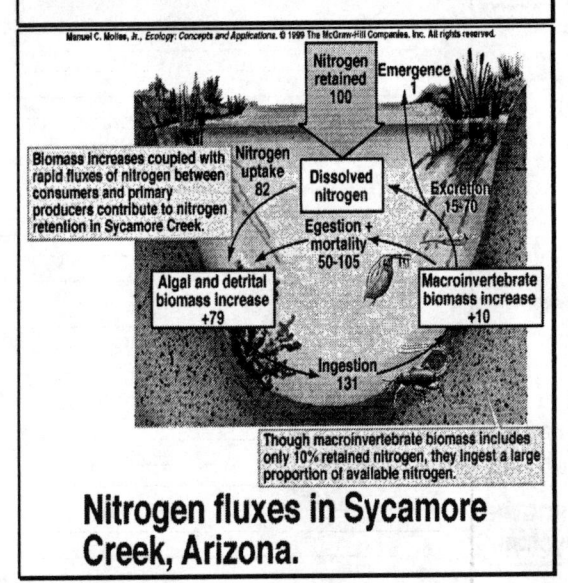

Biomass increases coupled with rapid fluxes of nitrogen between consumers and primary producers contribute to nitrogen retention in Sycamore Creek.

Nitrogen retained 100
Emergence 1
Nitrogen uptake 82
Dissolved nitrogen
Excretion 15-70
Egestion + mortality 50-105
Algal and detrital biomass increase +79
Macroinvertebrate biomass increase +10
Ingestion 131

Though macroinvertebrate biomass includes only 10% retained nitrogen, they ingest a large proportion of available nitrogen.

Nitrogen fluxes in Sycamore Creek, Arizona.

Figure 16.15

Pocket gophers and ecosystem structure.

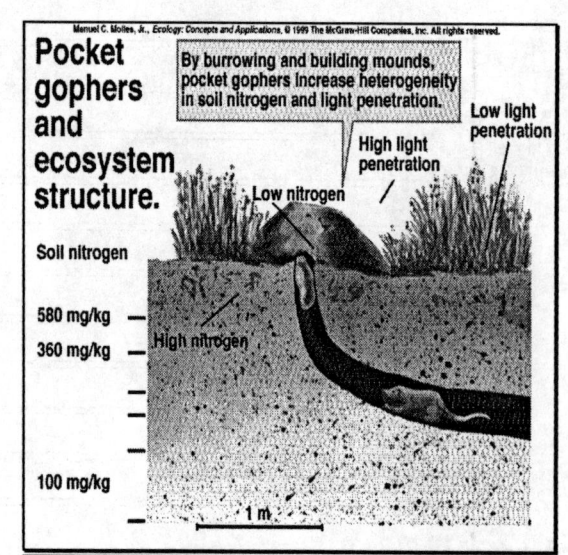

Figure 16.16

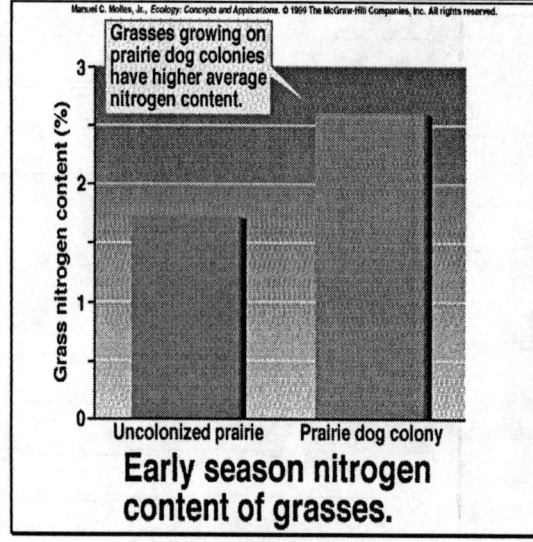

Early season nitrogen content of grasses.

Figure 16.17

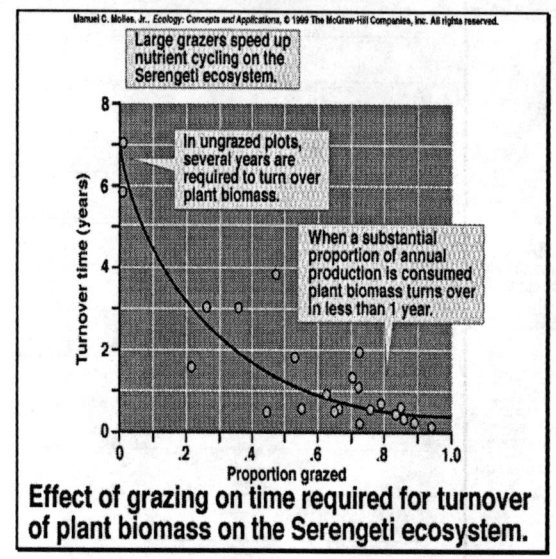

Effect of grazing on time required for turnover of plant biomass on the Serengeti ecosystem.

Figure 16.18

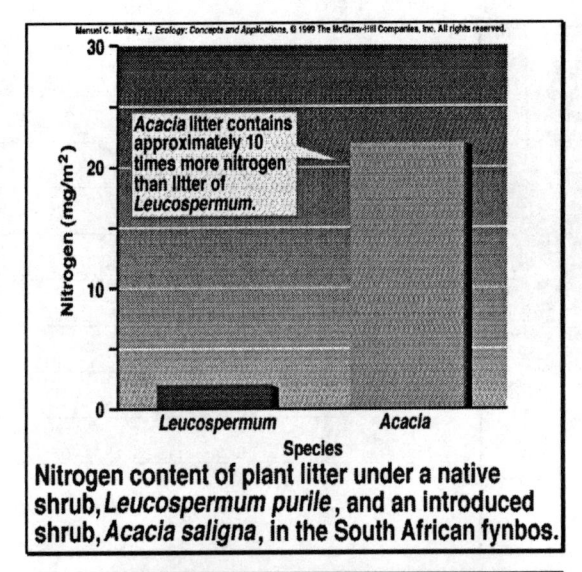

Nitrogen content of plant litter under a native shrub, *Leucospermum purile*, and an introduced shrub, *Acacia saligna*, in the South African fynbos.

Acacia litter contains approximately 10 times more nitrogen than litter of *Leucospermum*.

Figure 16.19

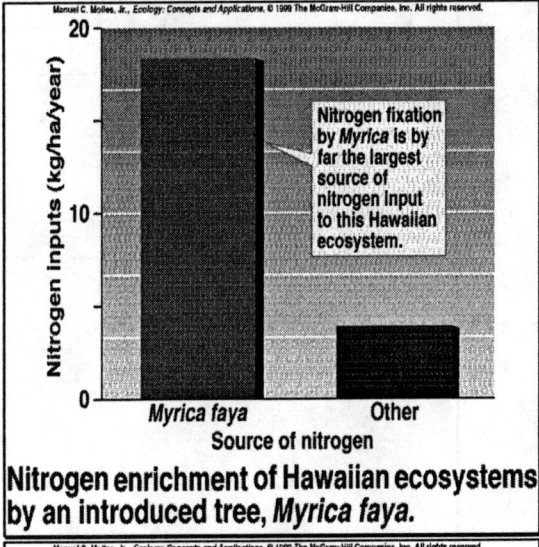

Nitrogen enrichment of Hawaiian ecosystems by an introduced tree, *Myrica faya*.

Nitrogen fixation by *Myrica* is by far the largest source of nitrogen input to this Hawaiian ecosystem.

Figure 16.20

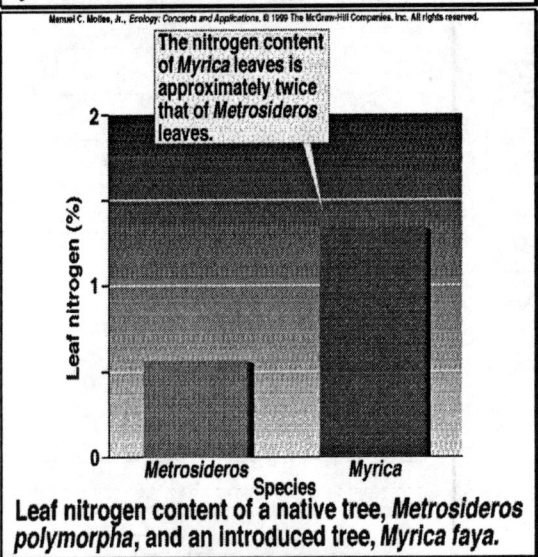

Leaf nitrogen content of a native tree, *Metrosideros polymorpha*, and an introduced tree, *Myrica faya*.

The nitrogen content of *Myrica* leaves is approximately twice that of *Metrosideros* leaves.

Figure 16.21

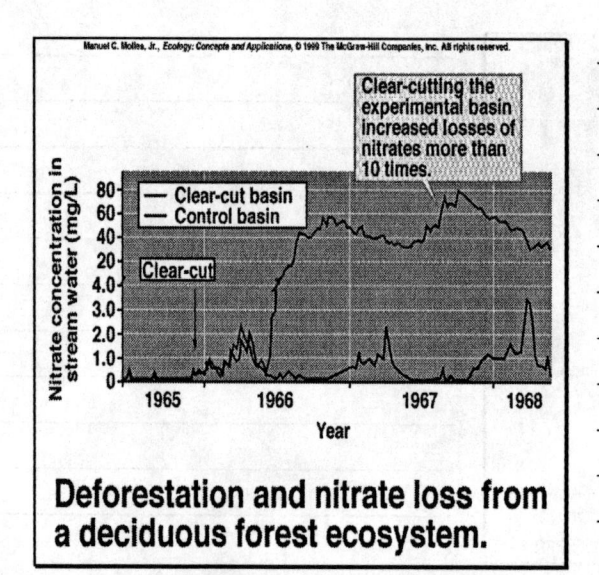

Deforestation and nitrate loss from a deciduous forest ecosystem.

Figure 16.22

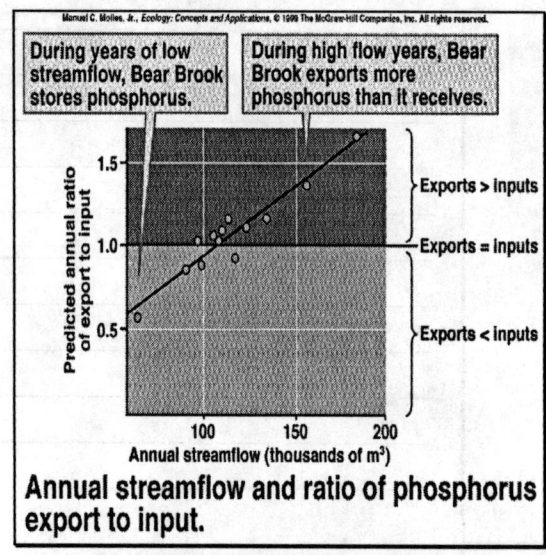

Annual streamflow and ratio of phosphorus export to input.

Figure 16.23

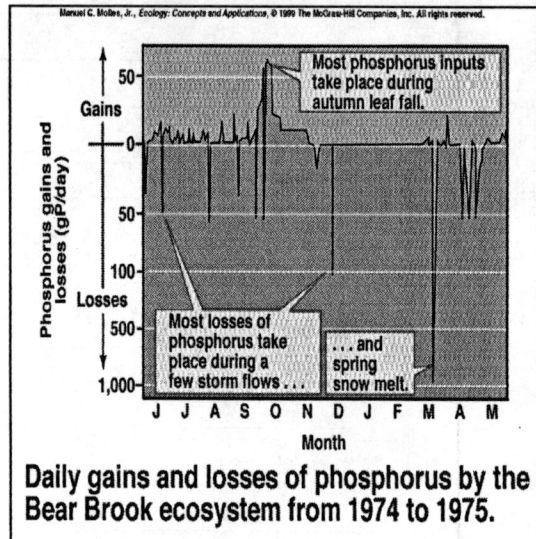

Daily gains and losses of phosphorus by the Bear Brook ecosystem from 1974 to 1975.

Figure 16.24

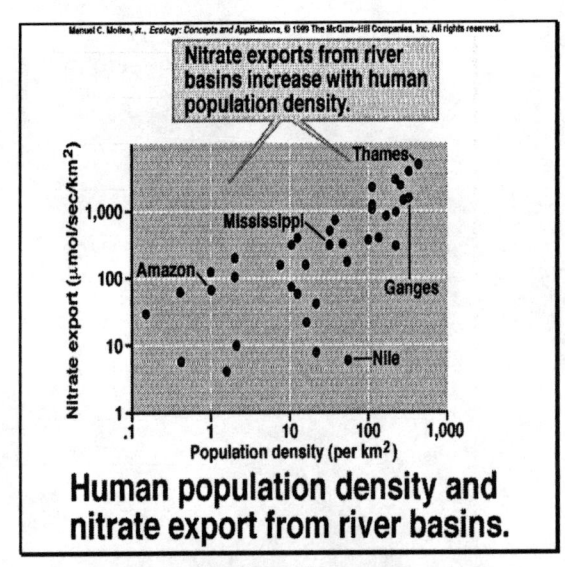

Human population density and nitrate export from river basins.

Figure 16.25

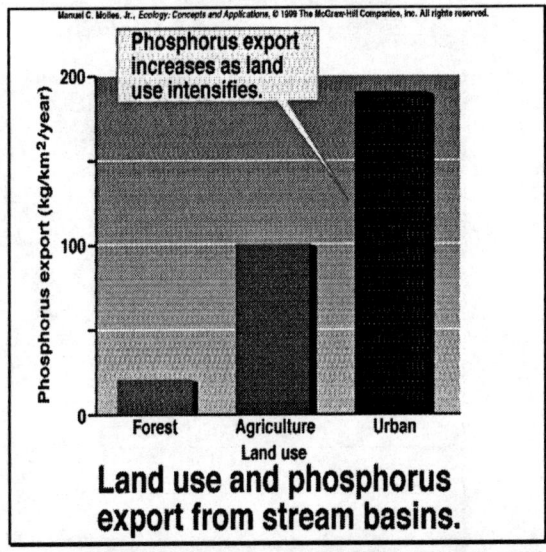

Land use and phosphorus export from stream basins.

Figure 16.26

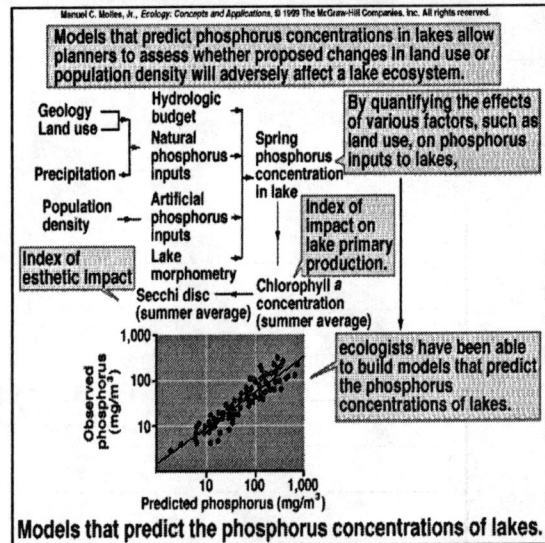

Models that predict the phosphorus concentrations of lakes.

Chapter 17

Ecology
Concepts and Applications
First Edition

Manuel C. Molles, Jr.

(c) The McGraw-Hill Companies, Inc.

Figure 17.2

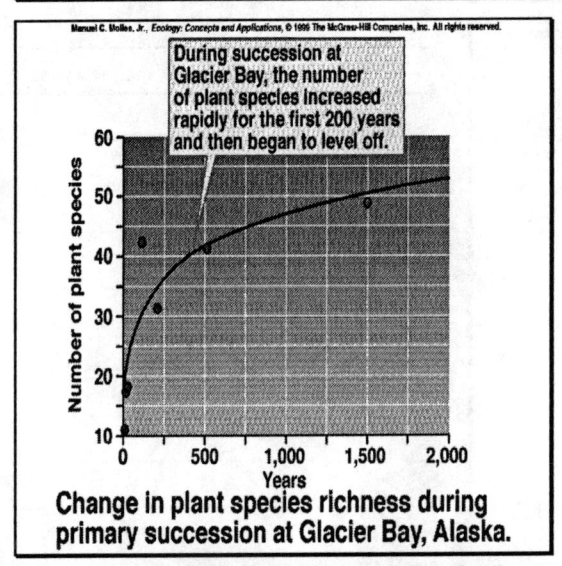

During succession at Glacier Bay, the number of plant species increased rapidly for the first 200 years and then began to level off.

Change in plant species richness during primary succession at Glacier Bay, Alaska.

Figure 17.3

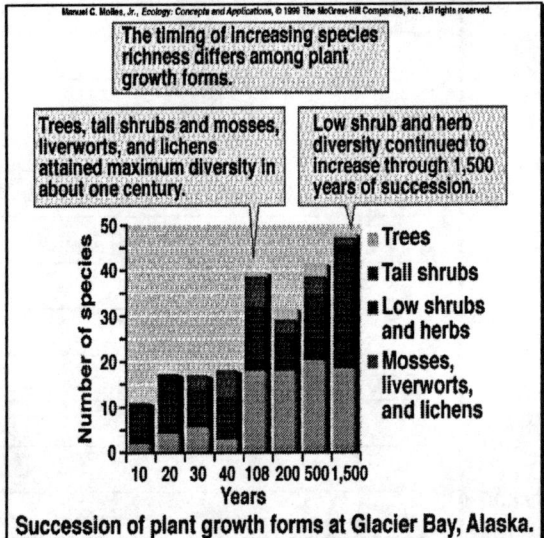

The timing of increasing species richness differs among plant growth forms.

Trees, tall shrubs and mosses, liverworts, and lichens attained maximum diversity in about one century.

Low shrub and herb diversity continued to increase through 1,500 years of succession.

Succession of plant growth forms at Glacier Bay, Alaska.

Figure 17.4

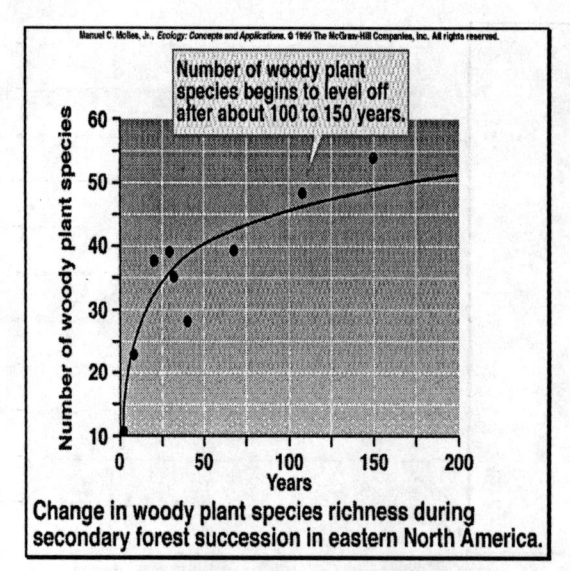

Number of woody plant species begins to level off after about 100 to 150 years.

Change in woody plant species richness during secondary forest succession in eastern North America.

Figure 17.5

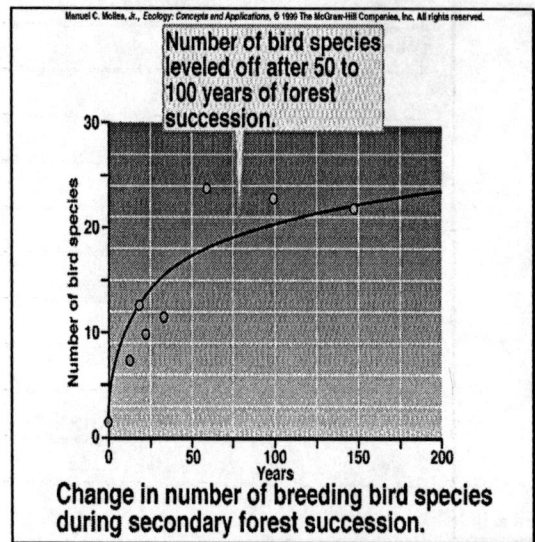

Number of bird species leveled off after 50 to 100 years of forest succession.

Change in number of breeding bird species during secondary forest succession.

Figure 17.7

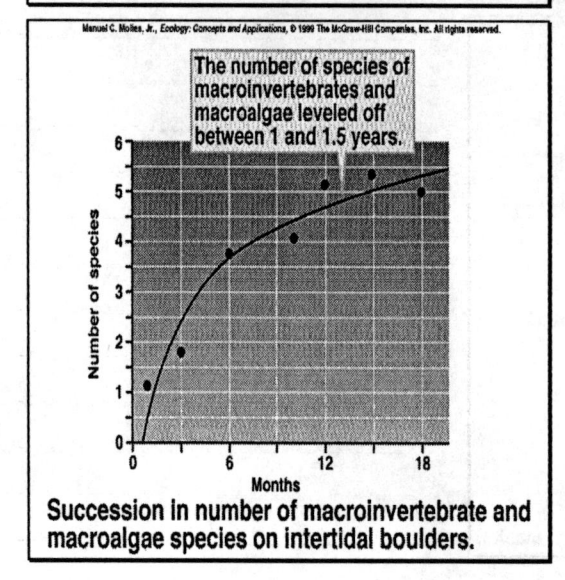

The number of species of macroinvertebrates and macroalgae leveled off between 1 and 1.5 years.

Succession in number of macroinvertebrate and macroalgae species on intertidal boulders.

Figure 17.8

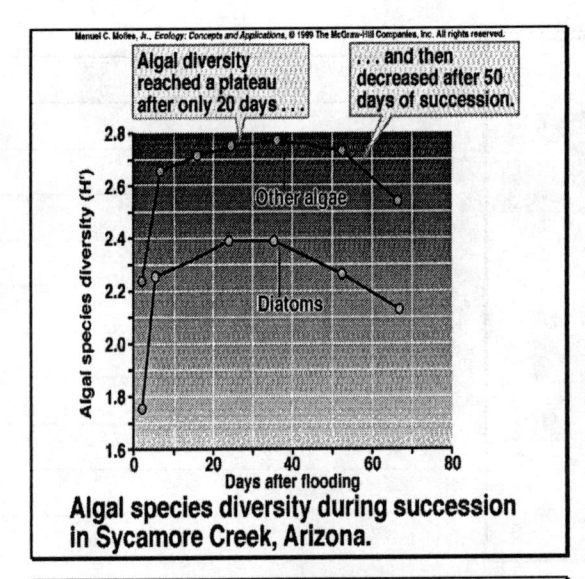

Algal diversity reached a plateau after only 20 days... ...and then decreased after 50 days of succession.

Algal species diversity during succession in Sycamore Creek, Arizona.

Figure 17.9

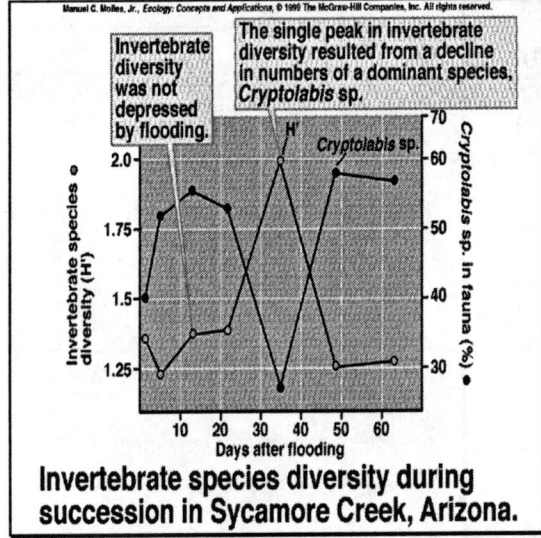

Invertebrate diversity was not depressed by flooding. The single peak in invertebrate diversity resulted from a decline in numbers of a dominant species, Cryptolabis sp.

Invertebrate species diversity during succession in Sycamore Creek, Arizona.

Figure 17.10

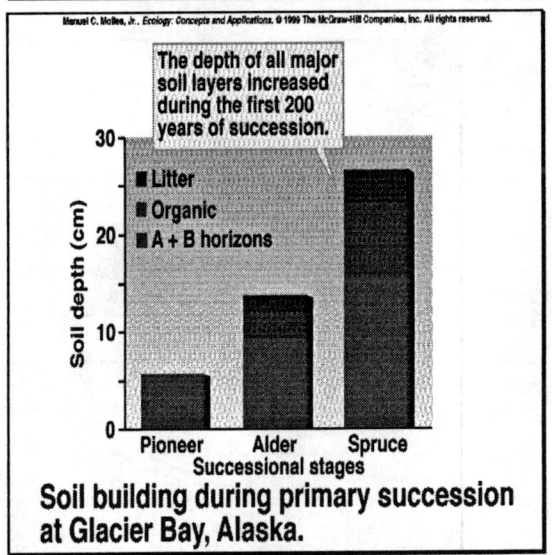

The depth of all major soil layers increased during the first 200 years of succession.

Soil building during primary succession at Glacier Bay, Alaska.

Figure 17.11

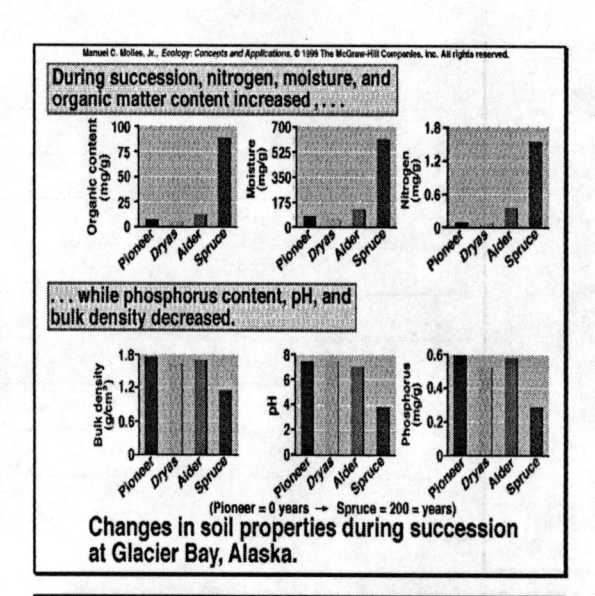

During succession, nitrogen, moisture, and organic matter content increased,...

...while phosphorus content, pH, and bulk density decreased.

(Pioneer = 0 years → Spruce = 200 = years)

Changes in soil properties during succession at Glacier Bay, Alaska.

Figure 17.12

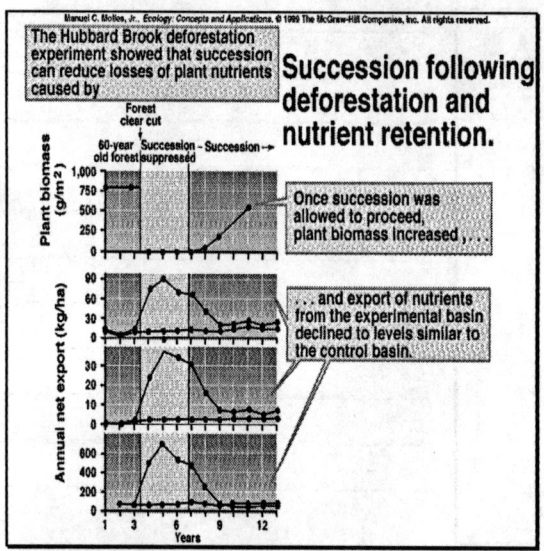

The Hubbard Brook deforestation experiment showed that succession can reduce losses of plant nutrients caused by

Succession following deforestation and nutrient retention.

Once succession was allowed to proceed, plant biomass increased,...

...and export of nutrients from the experimental basin declined to levels similar to the control basin.

Figure 17.13

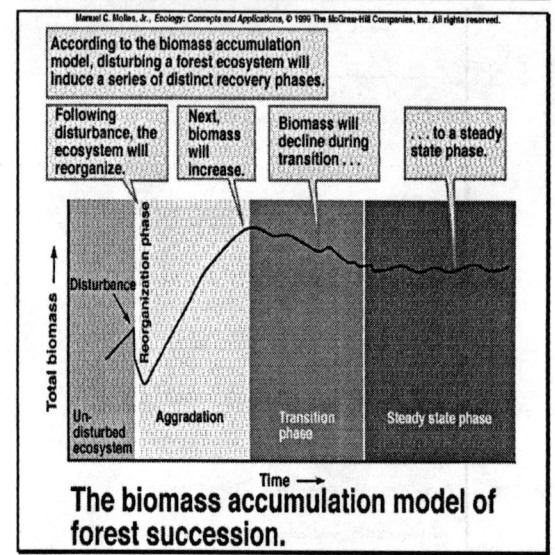

According to the biomass accumulation model, disturbing a forest ecosystem will induce a series of distinct recovery phases.

Following disturbance, the ecosystem will reorganize.

Next, biomass will increase.

Biomass will decline during transition...

...to a steady state phase.

The biomass accumulation model of forest succession.

Figure 17.14

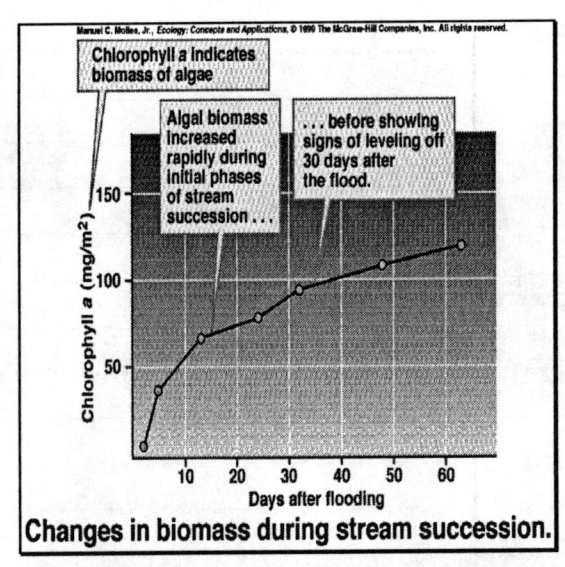

Changes in biomass during stream succession.
- Chlorophyll a indicates biomass of algae
- Algal biomass increased rapidly during initial phases of stream succession...
- ...before showing signs of leveling off 30 days after the flood.

Figure 17.15

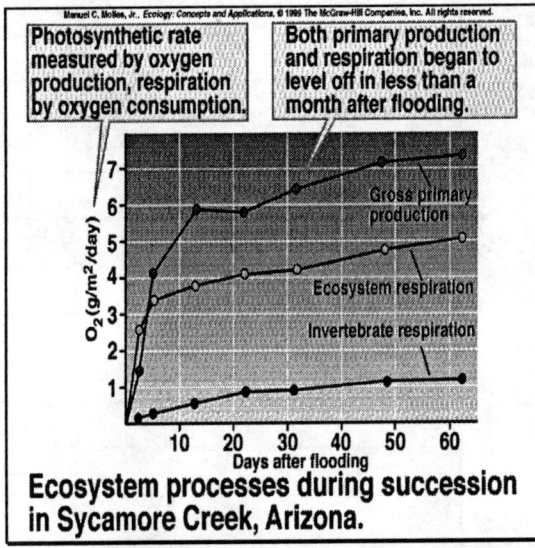

Ecosystem processes during succession in Sycamore Creek, Arizona.
- Photosynthetic rate measured by oxygen production, respiration by oxygen consumption.
- Both primary production and respiration began to level off in less than a month after flooding.
- Gross primary production
- Ecosystem respiration
- Invertebrate respiration

Figure 17.16

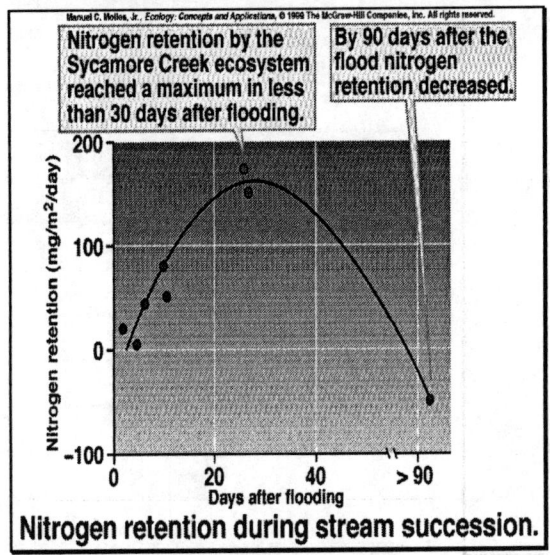

Nitrogen retention during stream succession.
- Nitrogen retention by the Sycamore Creek ecosystem reached a maximum in less than 30 days after flooding.
- By 90 days after the flood nitrogen retention decreased.

Figure 17.17

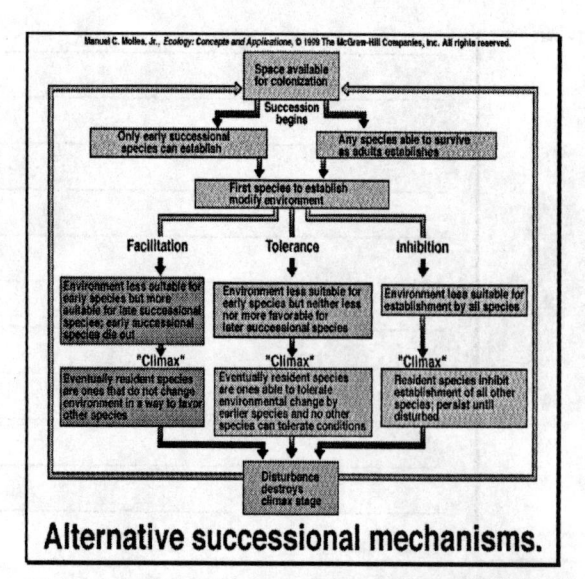

Alternative successional mechanisms.

Figure 17.18

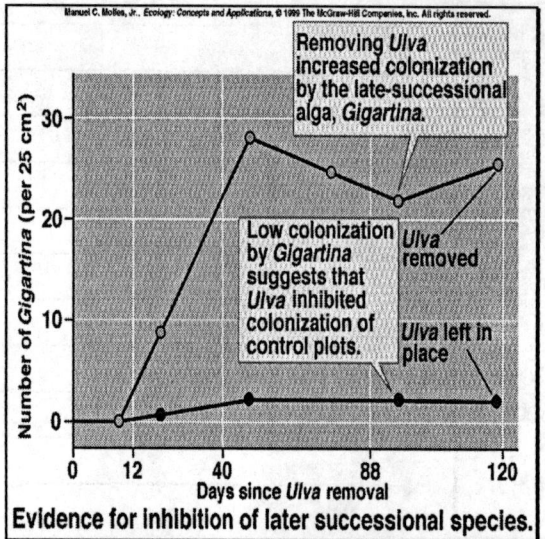

Evidence for inhibition of later successional species.

Figure 17.19

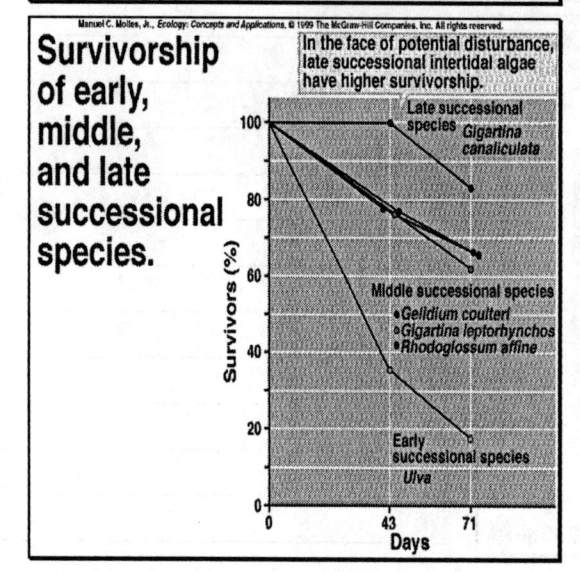

Survivorship of early, middle, and late successional species.

Figure 17.20

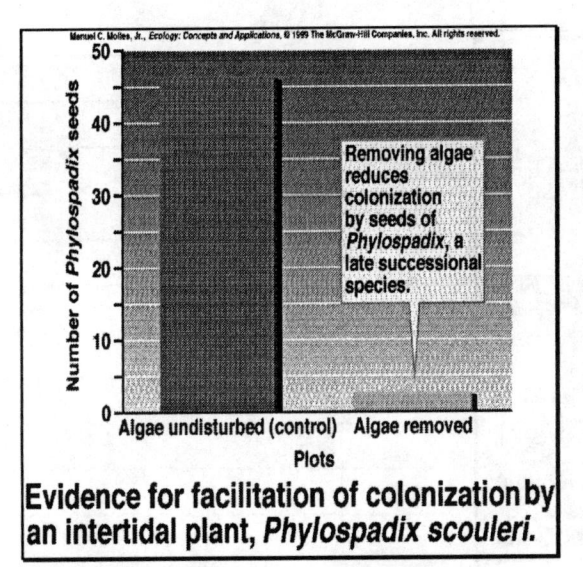

Evidence for facilitation of colonization by an intertidal plant, *Phylospadix scouleri*.

Removing algae reduces colonization by seeds of *Phylospadix*, a late successional species.

(Bar graph: Number of *Phylospadix* seeds vs. Plots — Algae undisturbed (control) ~46; Algae removed ~0)

Figure 17.21

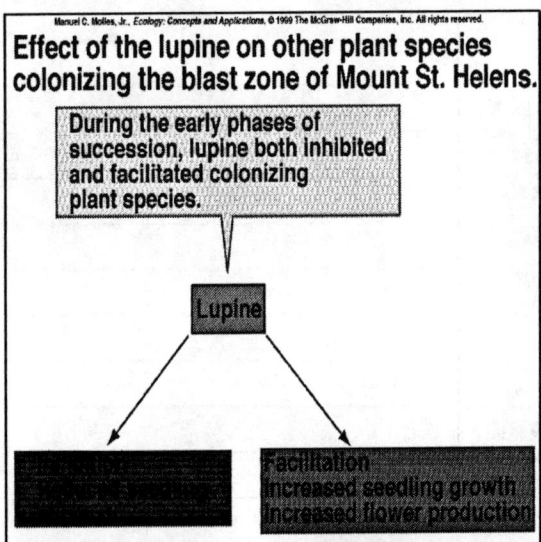

Effect of the lupine on other plant species colonizing the blast zone of Mount St. Helens.

During the early phases of succession, lupine both inhibited and facilitated colonizing plant species.

Lupine →
- Inhibition
- Facilitation: Increased seedling growth, Increased flower production

Figure 17.22

Inhibition and facilitation of spruce during the major successional stages at Glacier Bay, Alaska.

Pioneer	Dryas	Alder	Spruce
Inhibition / **Facilitation**	**Inhibition** / **Facilitation**	**Inhibition** / **Facilitation**	**Inhibition** / **Facilitation**
Lower germination / Higher survival	Lower germination / Higher nitrogen	Lower germination / Higher survival	Lower growth / Higher germination
Lower survival / Higher growth	Lower survival / Higher nitrogen	Lower survival	
Higher seed mortality	Higher seed mortality	More mycorrhizae	Higher seed mortality
	Root competition	Higher growth	Lower nitrogen
	Light competition		Root competition
			Light competition

Figure 17.23

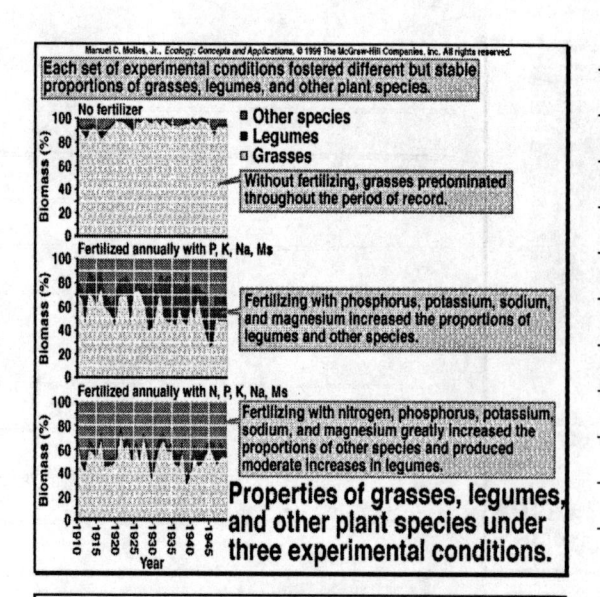

Properties of grasses, legumes, and other plant species under three experimental conditions.

Figure 17.24

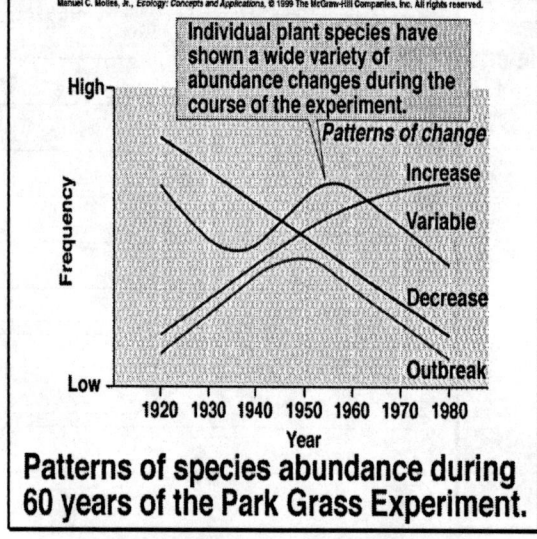

Patterns of species abundance during 60 years of the Park Grass Experiment.

Figure 17.25

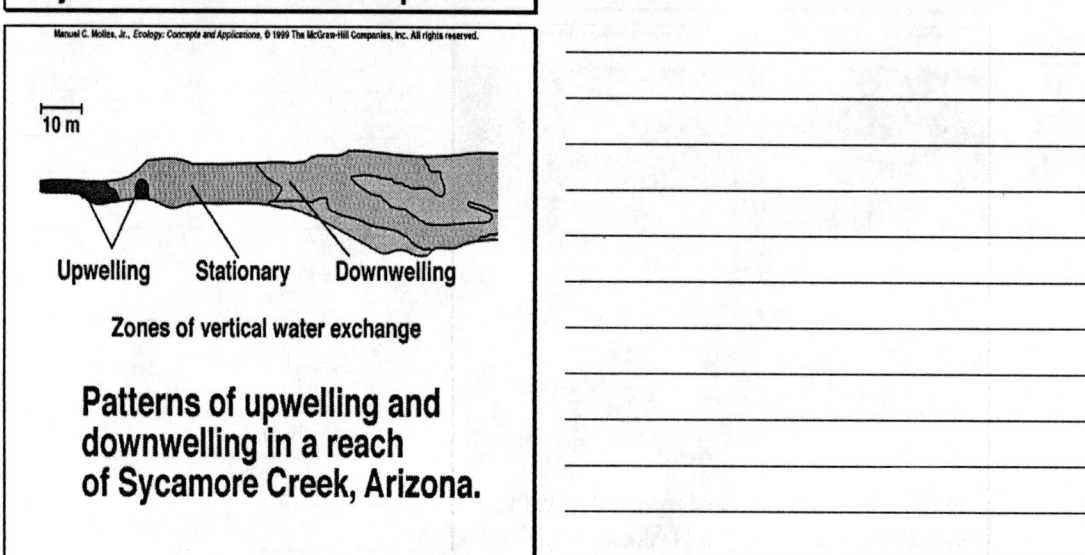

Patterns of upwelling and downwelling in a reach of Sycamore Creek, Arizona.

Figure 17.26

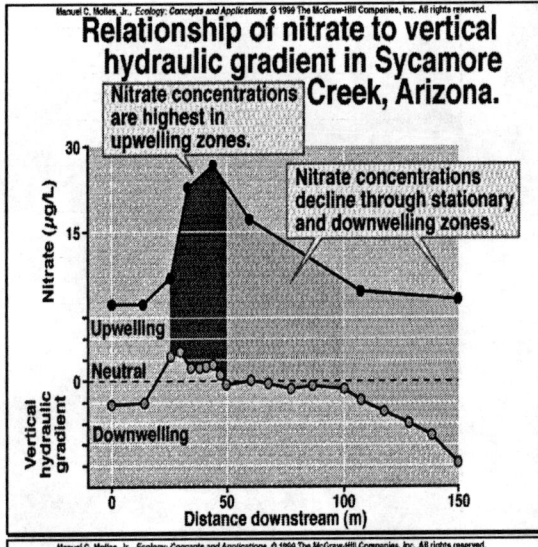

Relationship of nitrate to vertical hydraulic gradient in Sycamore Creek, Arizona.

Nitrate concentrations are highest in upwelling zones.

Nitrate concentrations decline through stationary and downwelling zones.

Figure 17.27

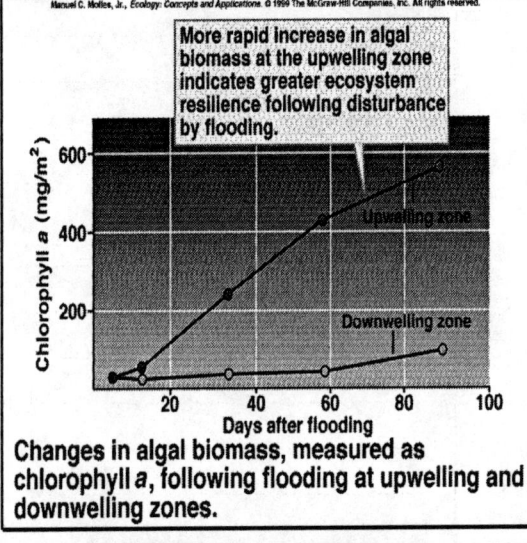

More rapid increase in algal biomass at the upwelling zone indicates greater ecosystem resilience following disturbance by flooding.

Changes in algal biomass, measured as chlorophyll *a*, following flooding at upwelling and downwelling zones.

Figure 17.30

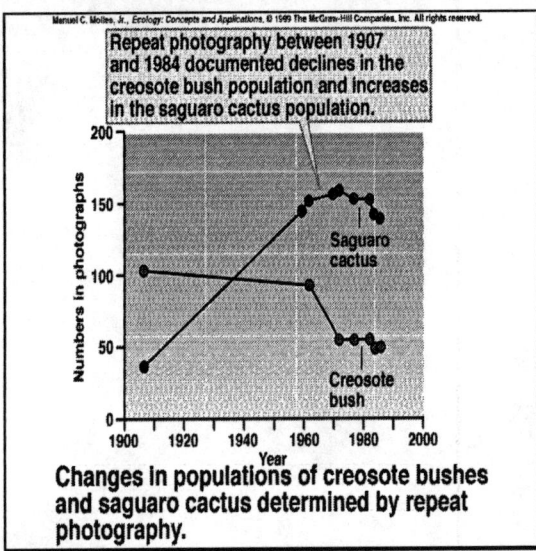

Repeat photography between 1907 and 1984 documented declines in the creosote bush population and increases in the saguaro cactus population.

Changes in populations of creosote bushes and saguaro cactus determined by repeat photography.

Chapter 18

Ecology

Concepts and Applications
First Edition

Manuel C. Molles, Jr.

(c) The McGraw-Hill Companies, Inc.

Figure 18.5

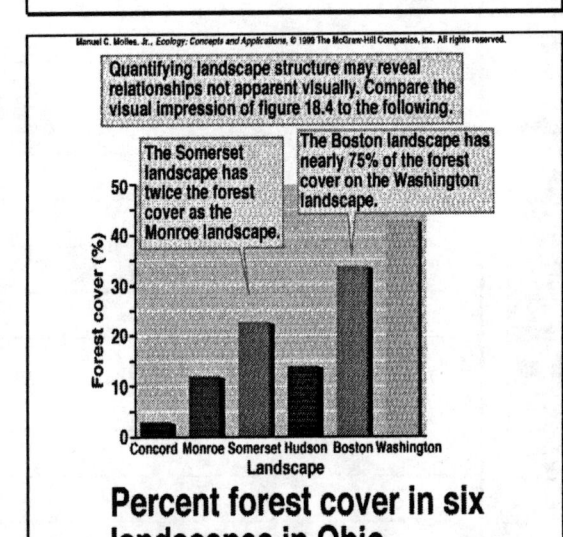

Quantifying landscape structure may reveal relationships not apparent visually. Compare the visual impression of figure 18.4 to the following.

The Somerset landscape has twice the forest cover as the Monroe landscape.

The Boston landscape has nearly 75% of the forest cover on the Washington landscape.

Percent forest cover in six landscapes in Ohio.

Figure 18.6

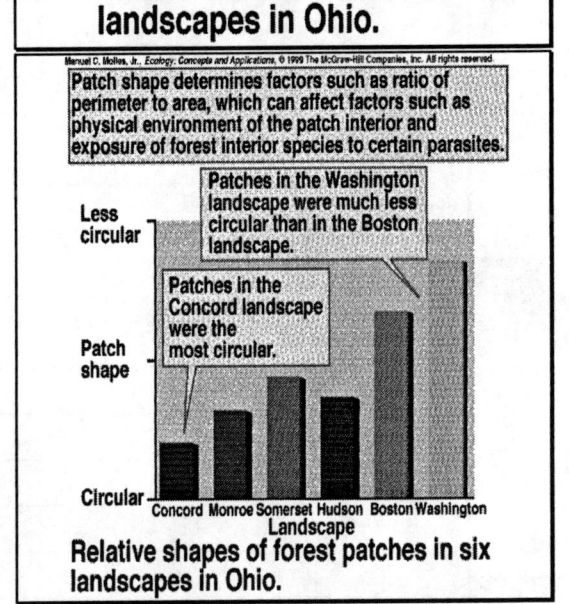

Patch shape determines factors such as ratio of perimeter to area, which can affect factors such as physical environment of the patch interior and exposure of forest interior species to certain parasites.

Patches in the Washington landscape were much less circular than in the Boston landscape.

Patches in the Concord landscape were the most circular.

Relative shapes of forest patches in six landscapes in Ohio.

Figure 18.7

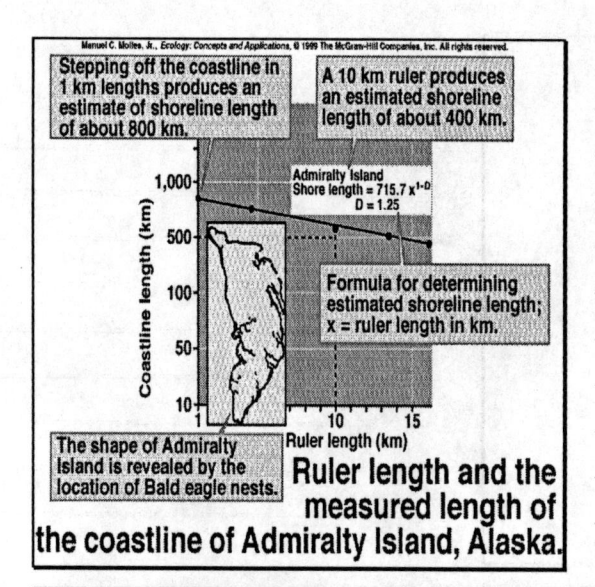

Figure 18.10

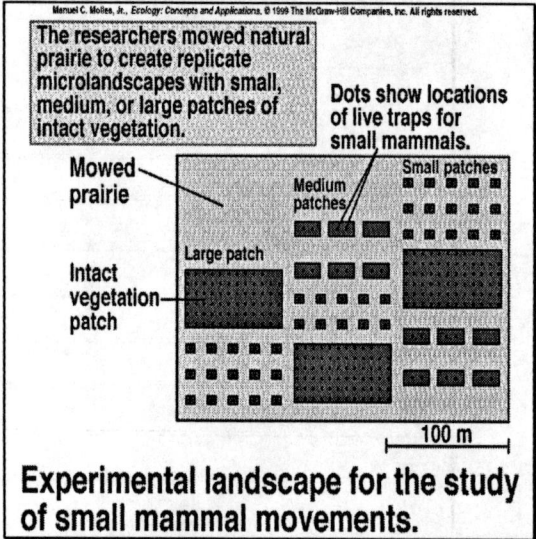

Figure 18.11

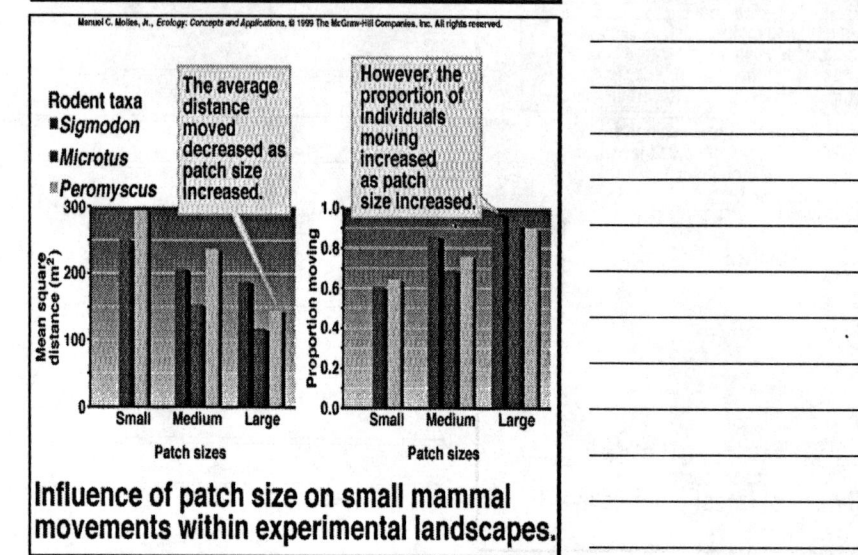

Figure 18.13

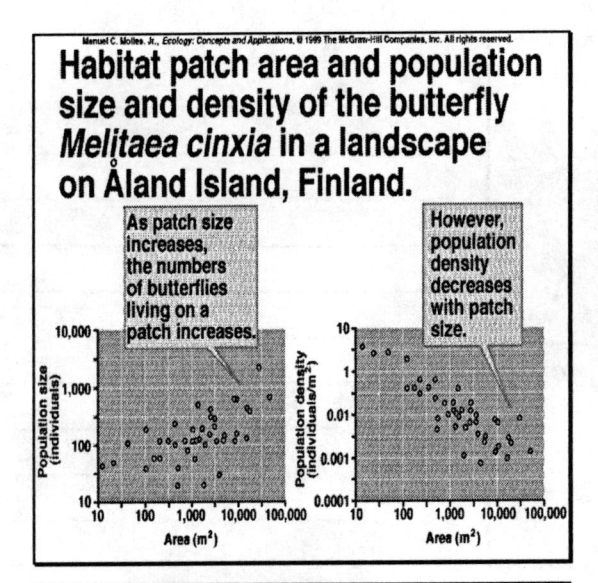

Habitat patch area and population size and density of the butterfly *Melitaea cinxia* in a landscape on Åland Island, Finland.

As patch size increases, the numbers of butterflies living on a patch increases.

However, population density decreases with patch size.

Figure 18.14

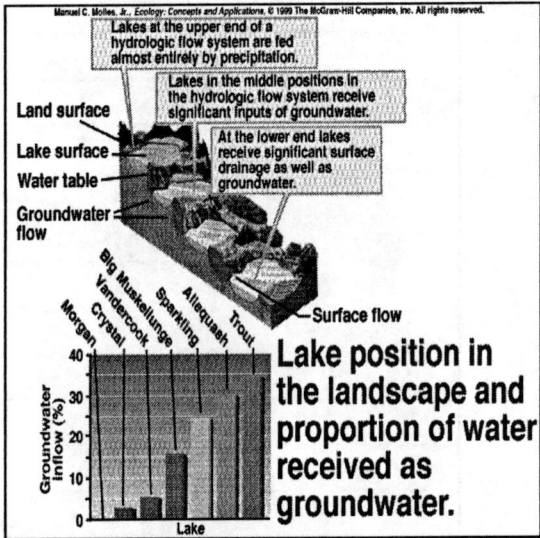

Lakes at the upper end of a hydrologic flow system are fed almost entirely by precipitation.

Lakes in the middle positions in the hydrologic flow system receive significant inputs of groundwater.

At the lower end lakes receive significant surface drainage as well as groundwater.

Lake position in the landscape and proportion of water received as groundwater.

Figure 18.15

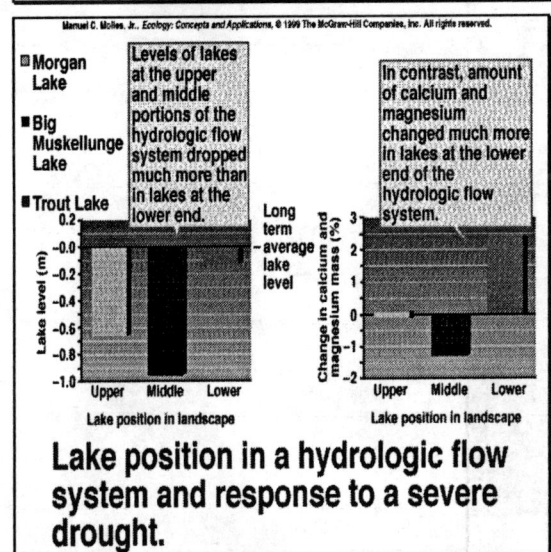

- Morgan Lake
- Big Muskellunge Lake
- Trout Lake

Levels of lakes at the upper and middle portions of the hydrologic flow system dropped much more than in lakes at the lower end.

In contrast, amount of calcium and magnesium changed much more in lakes at the lower end of the hydrologic flow system.

Lake position in a hydrologic flow system and response to a severe drought.

Figure 18.17

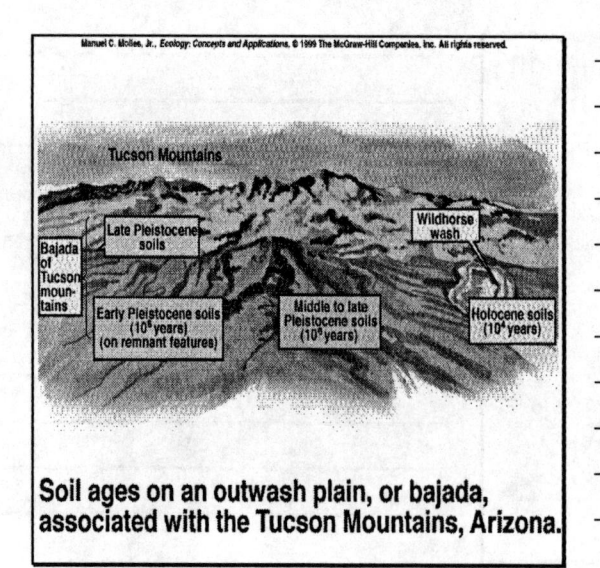

Soil ages on an outwash plain, or bajada, associated with the Tucson Mountains, Arizona.

Figure 18.18

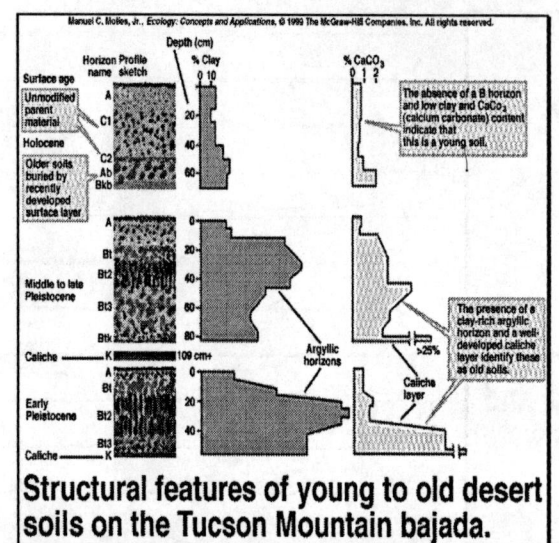

Structural features of young to old desert soils on the Tucson Mountain bajada.

Figure 18.19

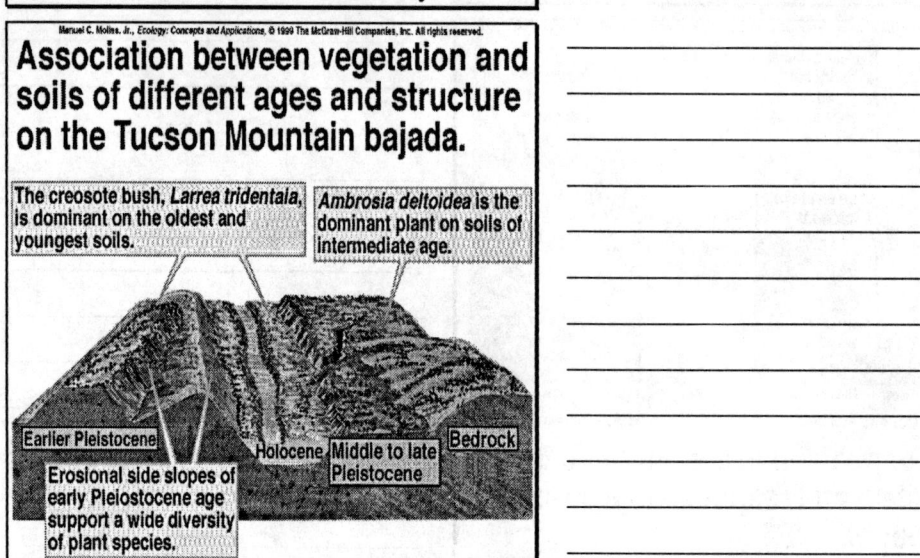

Association between vegetation and soils of different ages and structure on the Tucson Mountain bajada.

Figure 18.20

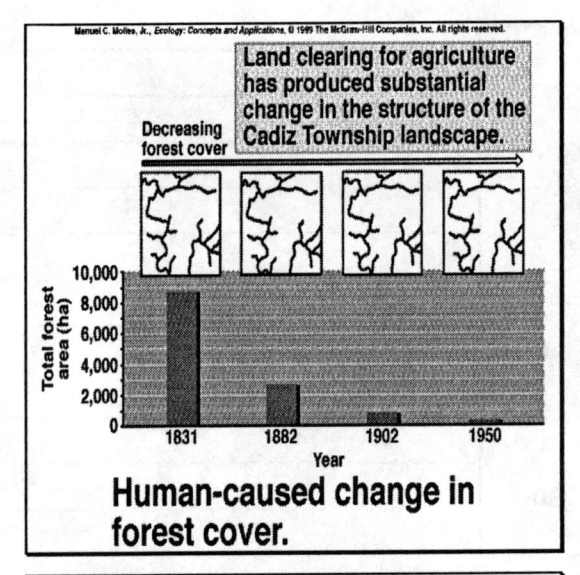

Land clearing for agriculture has produced substantial change in the structure of the Cadiz Township landscape.

Decreasing forest cover →

Human-caused change in forest cover.

Figure 18.21

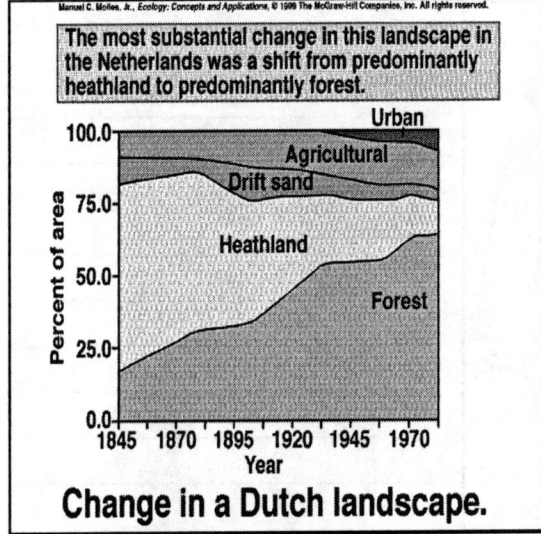

The most substantial change in this landscape in the Netherlands was a shift from predominantly heathland to predominantly forest.

Change in a Dutch landscape.

Figure 18.24

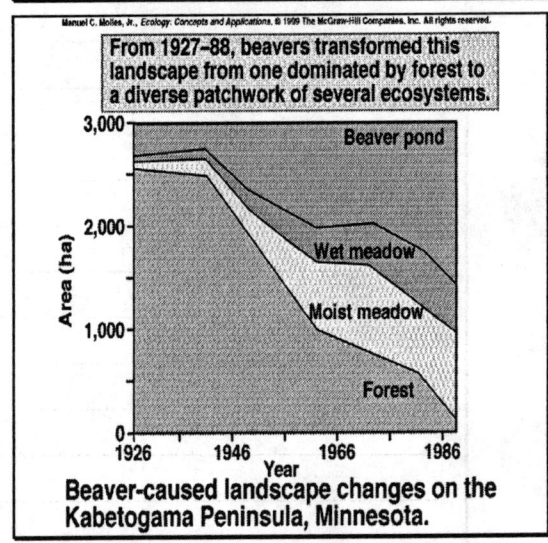

From 1927–88, beavers transformed this landscape from one dominated by forest to a diverse patchwork of several ecosystems.

Beaver-caused landscape changes on the Kabetogama Peninsula, Minnesota.

Figure 18.25

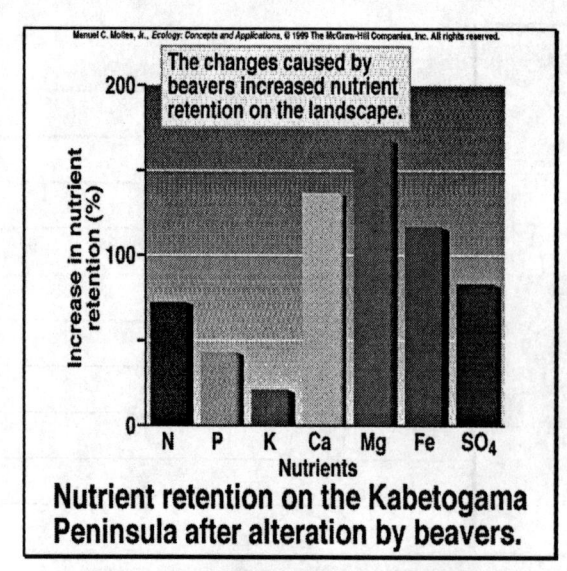

Nutrient retention on the Kabetogama Peninsula after alteration by beavers.

The changes caused by beavers increased nutrient retention on the landscape.

Figure 18.27

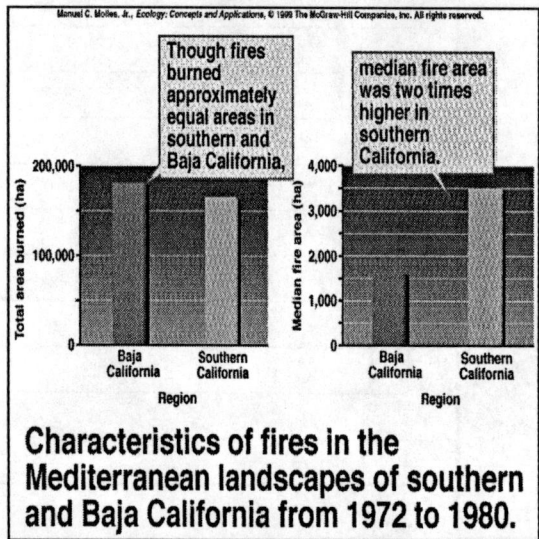

Characteristics of fires in the Mediterranean landscapes of southern and Baja California from 1972 to 1980.

Though fires burned approximately equal areas in southern and Baja California, median fire area was two times higher in southern California.

Review

Fragmented forest landscapes.

Landscape 1

Landscape 2

Review

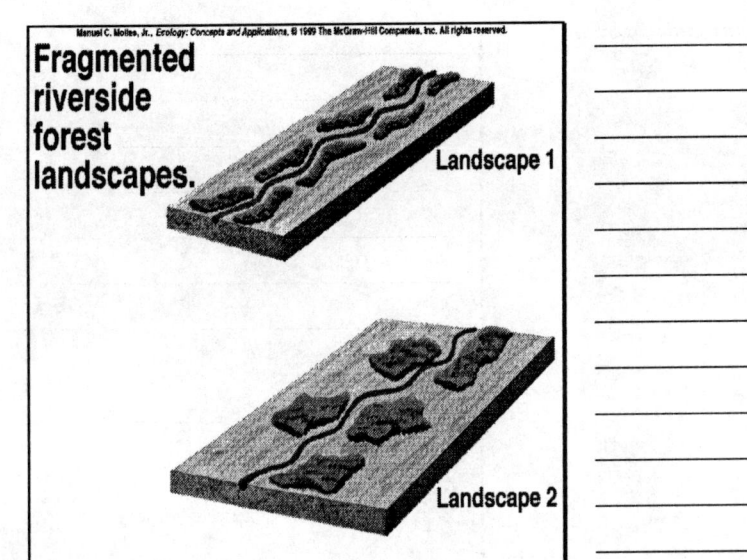

Fragmented riverside forest landscapes.

Landscape 1

Landscape 2

Chapter 19

Ecology
Concepts and Applications
First Edition

Manuel C. Molles, Jr.

(c) The McGraw-Hill Companies, Inc.

Figure 19.2

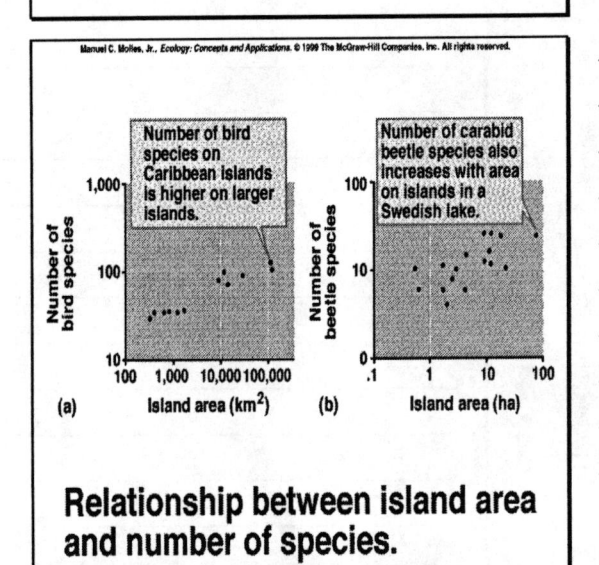

Relationship between island area and number of species.

(a) Number of bird species on Caribbean islands is higher on larger islands.

(b) Number of carabid beetle species also increases with area on islands in a Swedish lake.

Figure 19.3

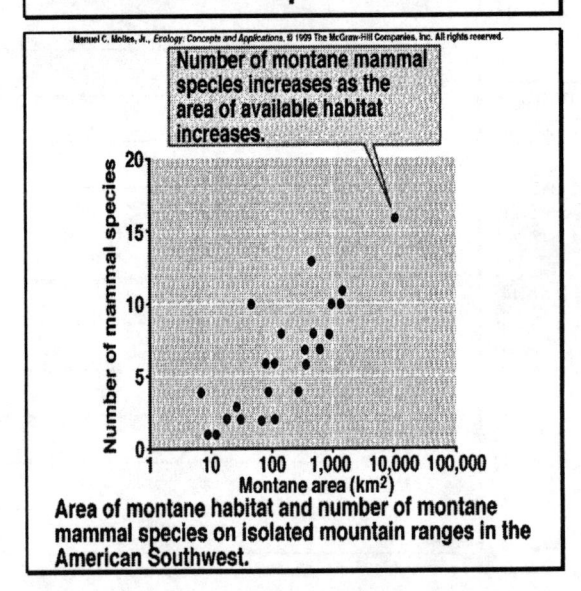

Number of montane mammal species increases as the area of available habitat increases.

Area of montane habitat and number of montane mammal species on isolated mountain ranges in the American Southwest.

Figure 19.4

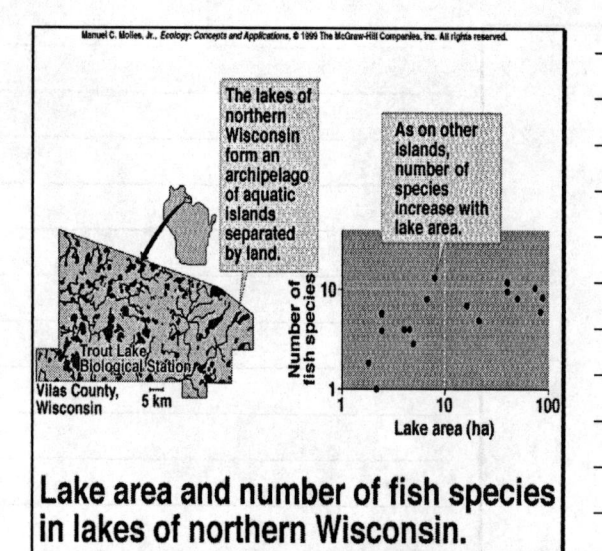

Lake area and number of fish species in lakes of northern Wisconsin.

Figure 19.5

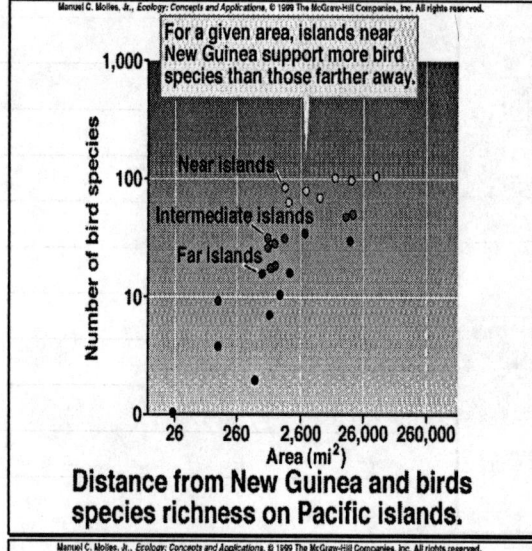

Distance from New Guinea and birds species richness on Pacific islands.

Figure 19.6

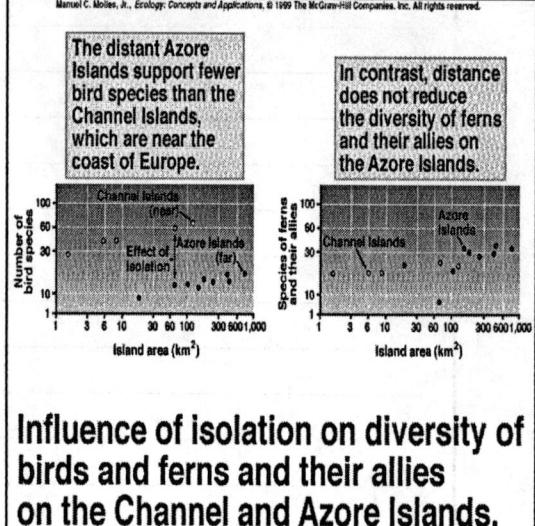

Influence of isolation on diversity of birds and ferns and their allies on the Channel and Azore Islands.

Figure 19.7

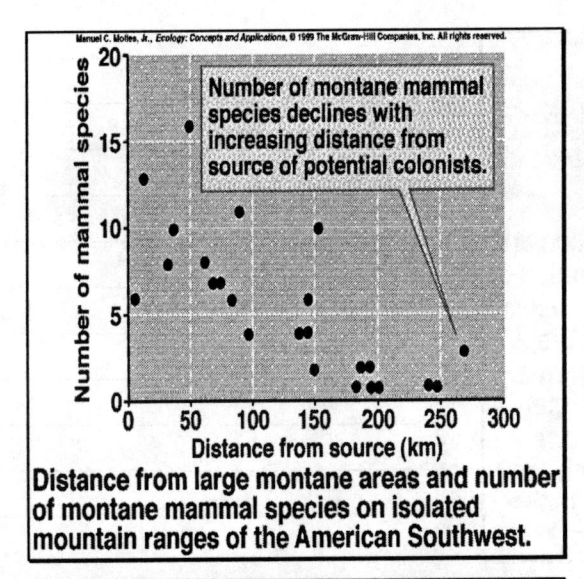

Distance from large montane areas and number of montane mammal species on isolated mountain ranges of the American Southwest.

Number of montane mammal species declines with increasing distance from source of potential colonists.

Figure 19.8

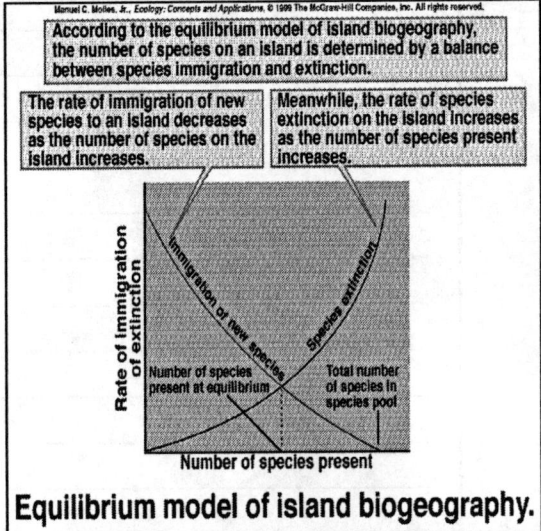

According to the equilibrium model of island biogeography, the number of species on an island is determined by a balance between species immigration and extinction.

The rate of immigration of new species to an island decreases as the number of species on the island increases.

Meanwhile, the rate of species extinction on the island increases as the number of species present increases.

Equilibrium model of island biogeography.

Figure 19.9

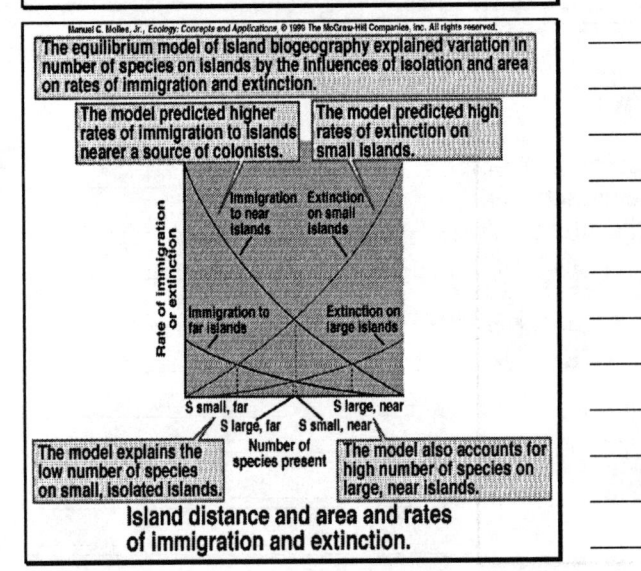

The equilibrium model of island biogeography explained variation in number of species on islands by the influences of isolation and area on rates of immigration and extinction.

The model predicted higher rates of immigration to islands nearer a source of colonists.

The model predicted high rates of extinction on small islands.

The model explains the low number of species on small, isolated islands.

The model also accounts for high number of species on large, near islands.

Island distance and area and rates of immigration and extinction.

Figure 19.10

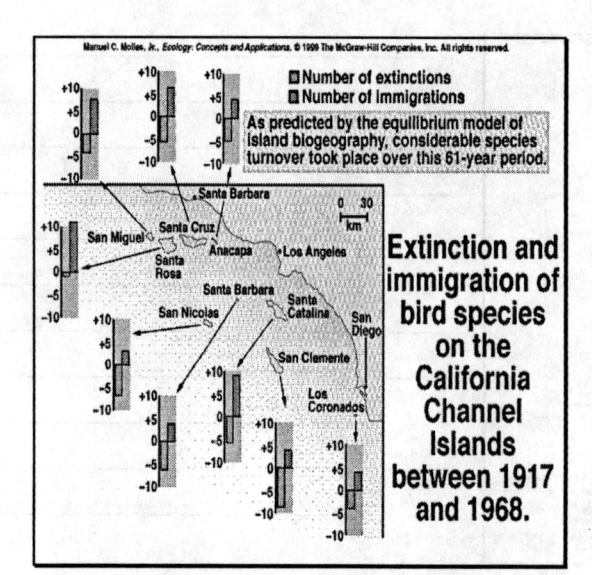

Extinction and immigration of bird species on the California Channel Islands between 1917 and 1968.

Figure 19.12

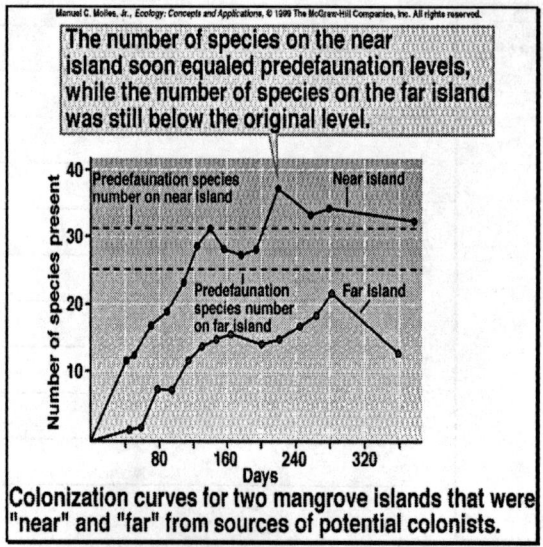

Colonization curves for two mangrove islands that were "near" and "far" from sources of potential colonists.

Figure 19.13

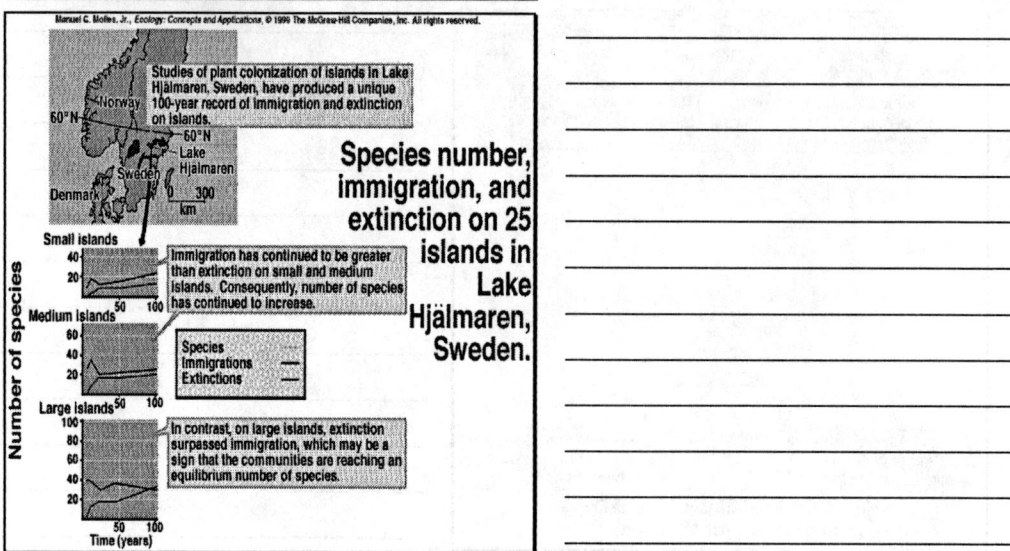

Species number, immigration, and extinction on 25 islands in Lake Hjälmaren, Sweden.

Figure 19.14

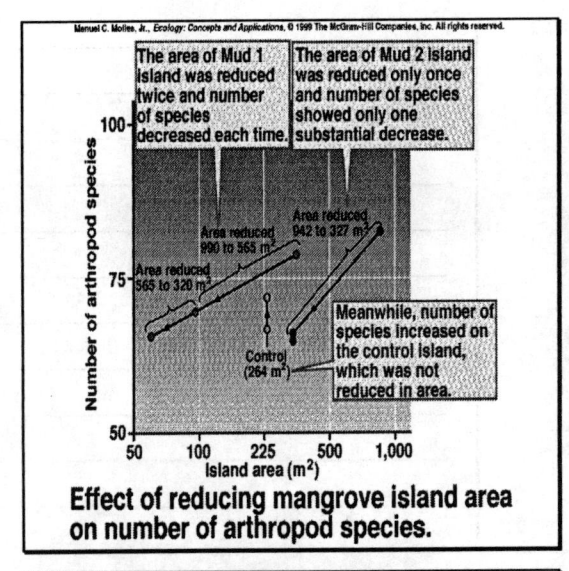

Effect of reducing mangrove island area on number of arthropod species.

Figure 19.15

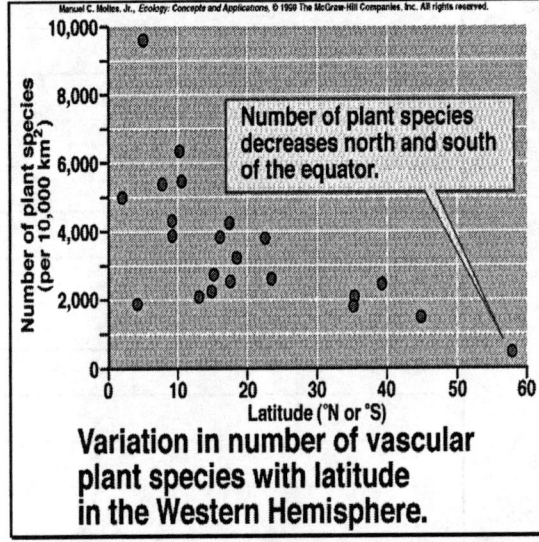

Variation in number of vascular plant species with latitude in the Western Hemisphere.

Figure 19.16

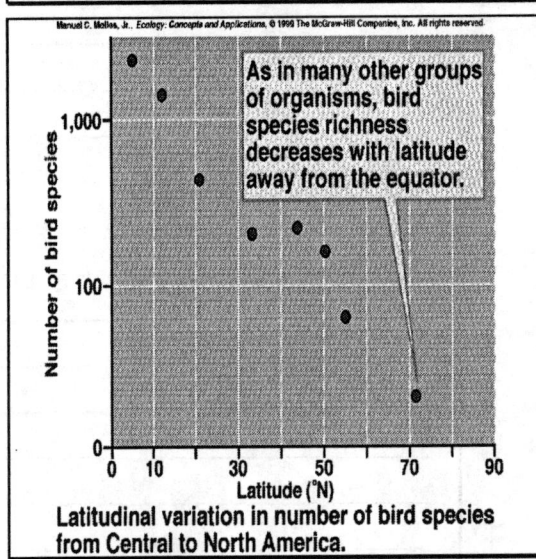

Latitudinal variation in number of bird species from Central to North America.

Figure 19.17

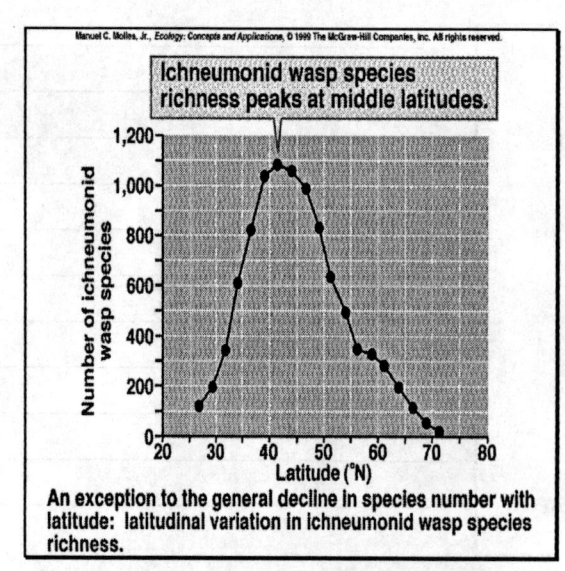

Ichneumonid wasp species richness peaks at middle latitudes.

An exception to the general decline in species number with latitude: latitudinal variation in ichneumonid wasp species richness.

Figure 19.18

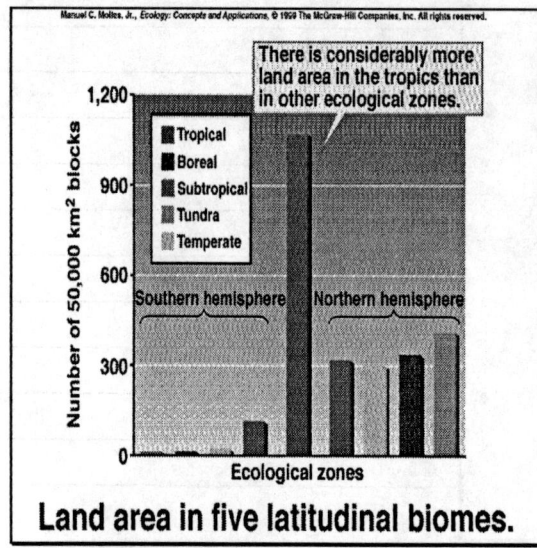

There is considerably more land area in the tropics than in other ecological zones.

Land area in five latitudinal biomes.

Figure 19.19

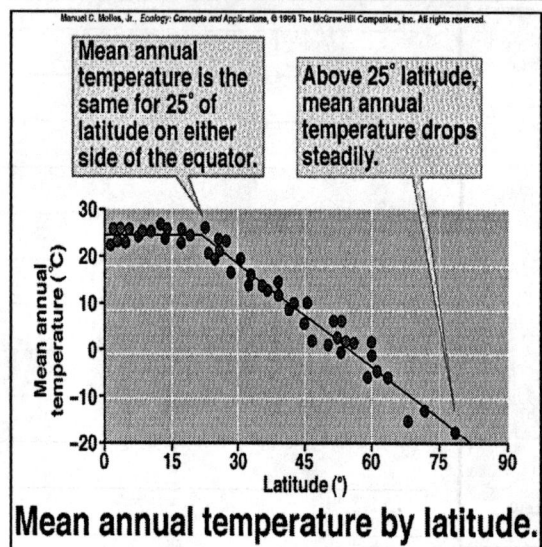

Mean annual temperature is the same for 25° of latitude on either side of the equator.

Above 25° latitude, mean annual temperature drops steadily.

Mean annual temperature by latitude.

Figure 19.20

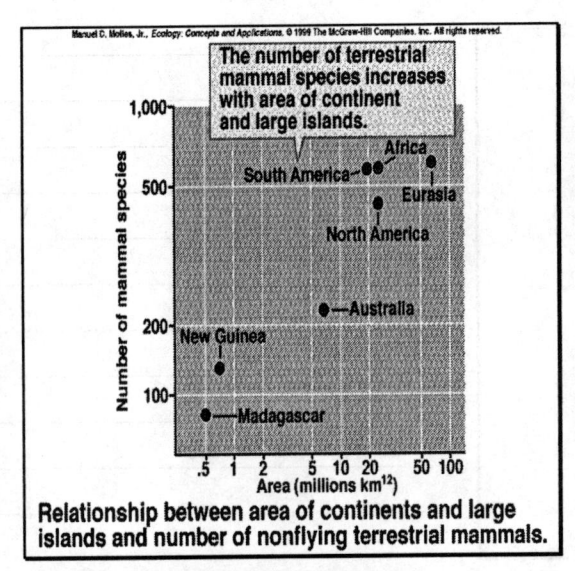

The number of terrestrial mammal species increases with area of continent and large islands.

Relationship between area of continents and large islands and number of nonflying terrestrial mammals.

Figure 19.21

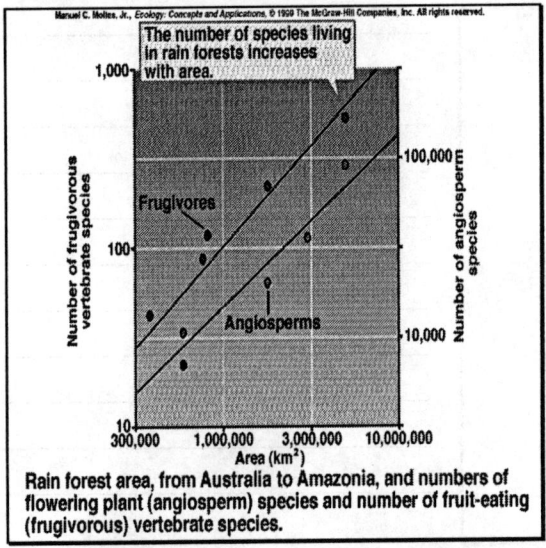

The number of species living in rain forests increases with area.

Rain forest area, from Australia to Amazonia, and numbers of flowering plant (angiosperm) species and number of fruit-eating (frugivorous) vertebrate species.

Figure 19.22

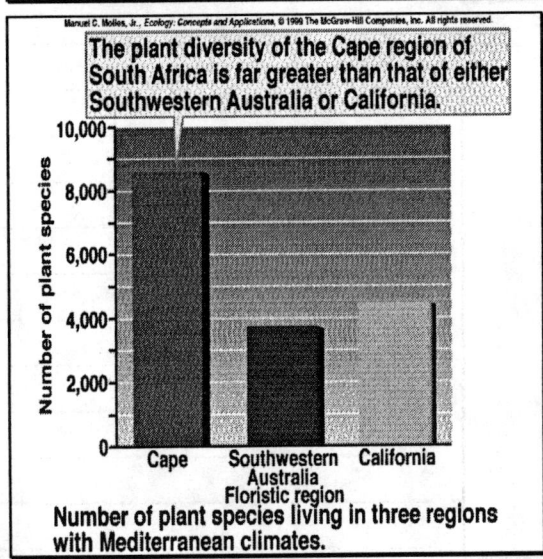

The plant diversity of the Cape region of South Africa is far greater than that of either Southwestern Australia or California.

Number of plant species living in three regions with Mediterranean climates.

Figure 19.23

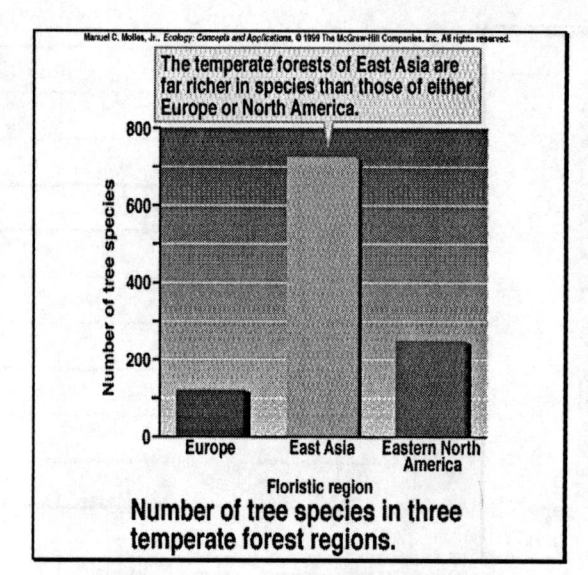

Number of tree species in three temperate forest regions.

The temperate forests of East Asia are far richer in species than those of either Europe or North America.

Figure 19.24

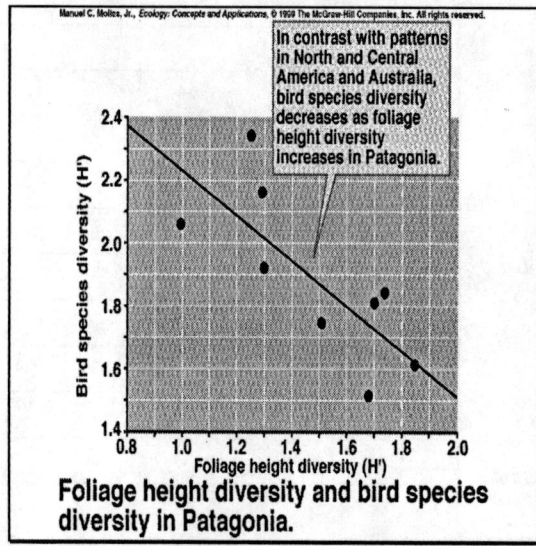

Foliage height diversity and bird species diversity in Patagonia.

In contrast with patterns in North and Central America and Australia, bird species diversity decreases as foliage height diversity increases in Patagonia.

Figure 19.25

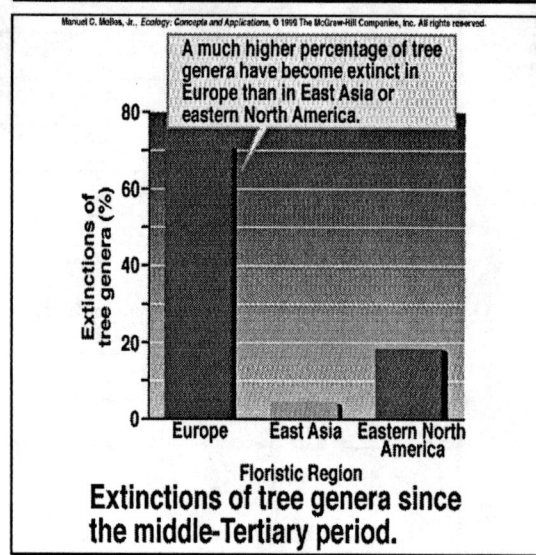

Extinctions of tree genera since the middle-Tertiary period.

A much higher percentage of tree genera have become extinct in Europe than in East Asia or eastern North America.

Figure 19.26

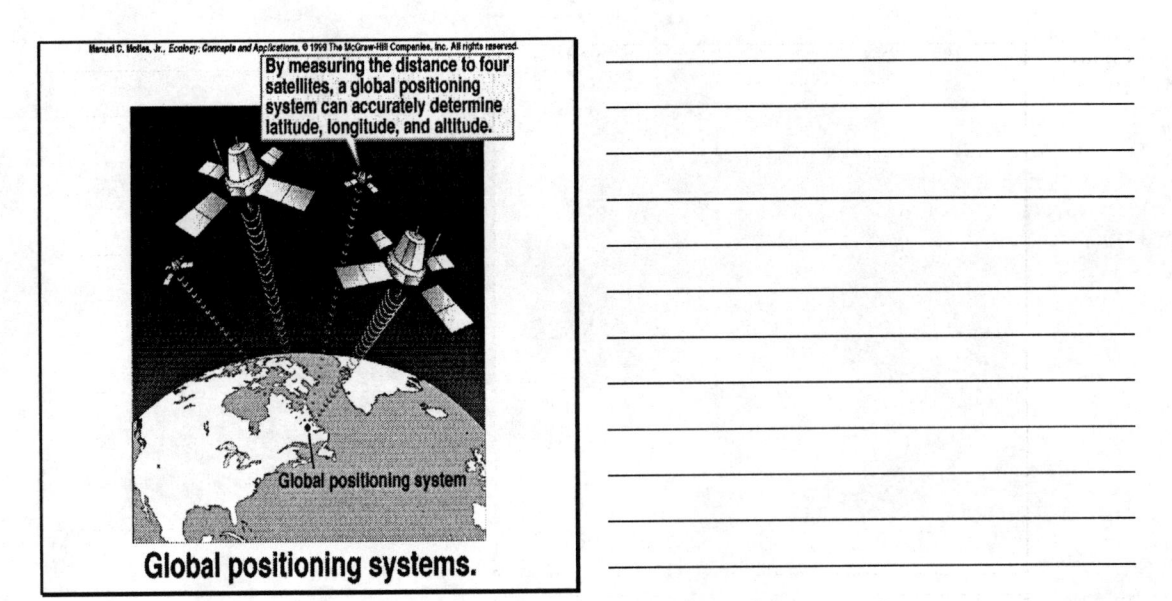

Global positioning systems.

By measuring the distance to four satellites, a global positioning system can accurately determine latitude, longitude, and altitude.

Chapter 20

Ecology

Concepts and Applications
First Edition

Manuel C. Molles, Jr.

(c) The McGraw-Hill Companies, Inc.

Figure 20.2

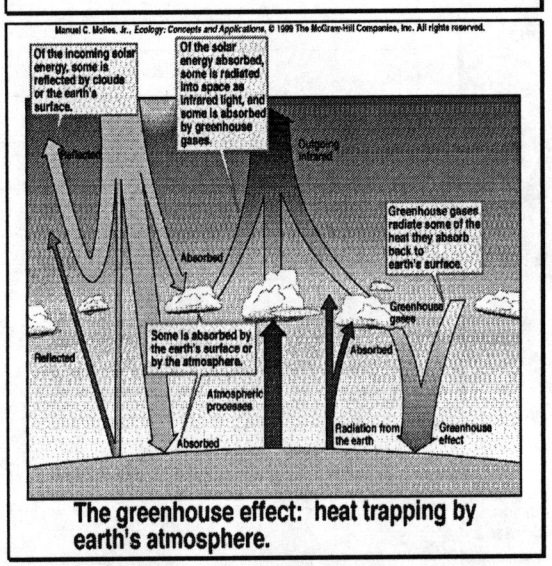

The greenhouse effect: heat trapping by earth's atmosphere.

- Of the incoming solar energy, some is reflected by clouds or the earth's surface.
- Of the solar energy absorbed, some is radiated into space as infrared light, and some is absorbed by greenhouse gases.
- Greenhouse gases radiate some of the heat they absorb back to earth's surface.
- Some is absorbed by the earth's surface or by the atmosphere.

Figure 20.3

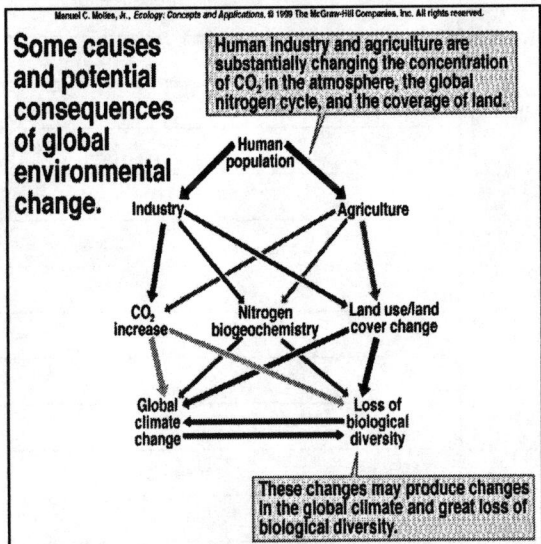

Some causes and potential consequences of global environmental change.

Human industry and agriculture are substantially changing the concentration of CO_2 in the atmosphere, the global nitrogen cycle, and the coverage of land.

Human population → Industry, Agriculture → CO_2 increase, Nitrogen biogeochemistry, Land use/land cover change → Global climate change, Loss of biological diversity

These changes may produce changes in the global climate and great loss of biological diversity.

167

Figure 20.4

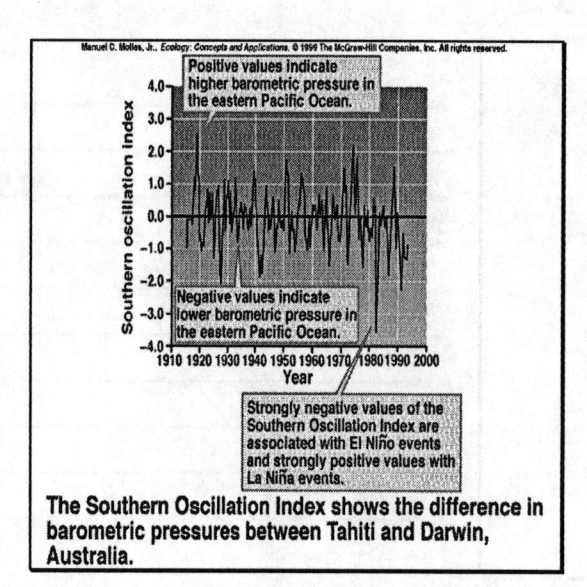

Positive values indicate higher barometric pressure in the eastern Pacific Ocean.

Negative values indicate lower barometric pressure in the eastern Pacific Ocean.

Strongly negative values of the Southern Oscillation Index are associated with El Niño events and strongly positive values with La Niña events.

The Southern Oscillation Index shows the difference in barometric pressures between Tahiti and Darwin, Australia.

Figure 20.5

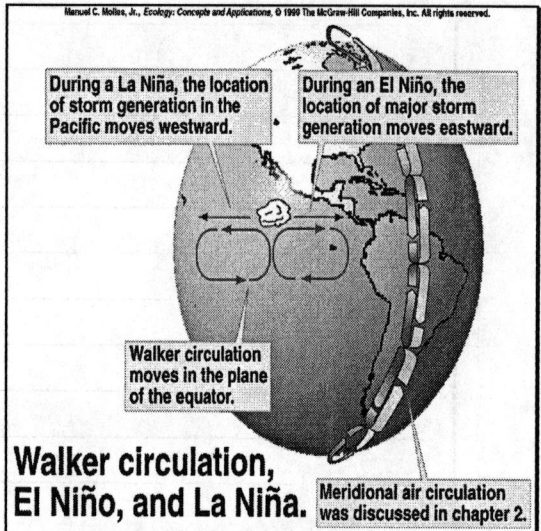

During a La Niña, the location of storm generation in the Pacific moves westward.

During an El Niño, the location of major storm generation moves eastward.

Walker circulation moves in the plane of the equator.

Meridional air circulation was discussed in chapter 2.

Walker circulation, El Niño, and La Niña.

Figure 20.6

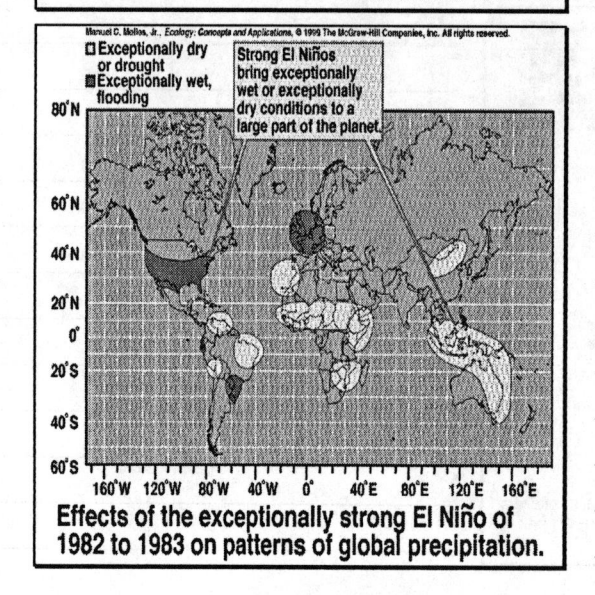

☐ Exceptionally dry or drought
▨ Exceptionally wet, flooding

Strong El Niños bring exceptionally wet or exceptionally dry conditions to a large part of the planet.

Effects of the exceptionally strong El Niño of 1982 to 1983 on patterns of global precipitation.

Figure 20.7

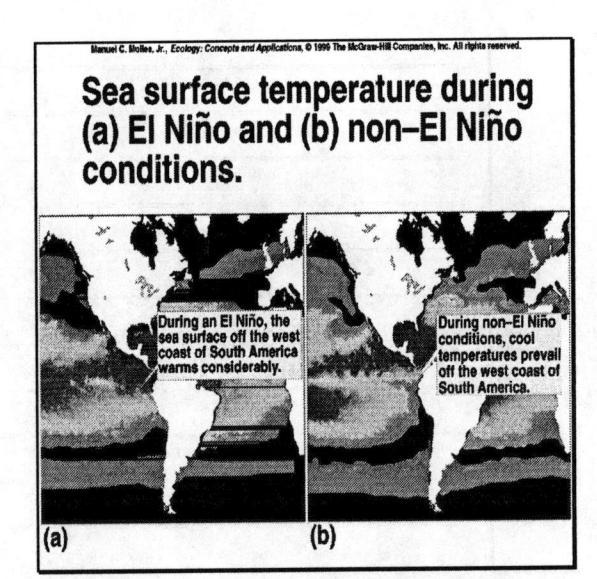

Sea surface temperature during (a) El Niño and (b) non–El Niño conditions.

Figure 20.10

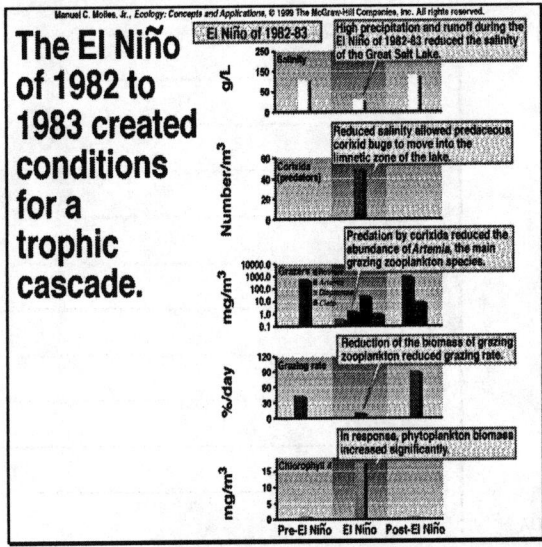

The El Niño of 1982 to 1983 created conditions for a trophic cascade.

Figure 20.12

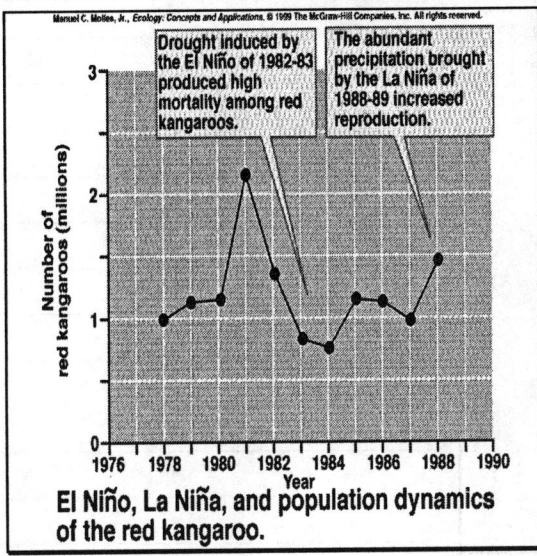

El Niño, La Niña, and population dynamics of the red kangaroo.

169

Figure 20.13

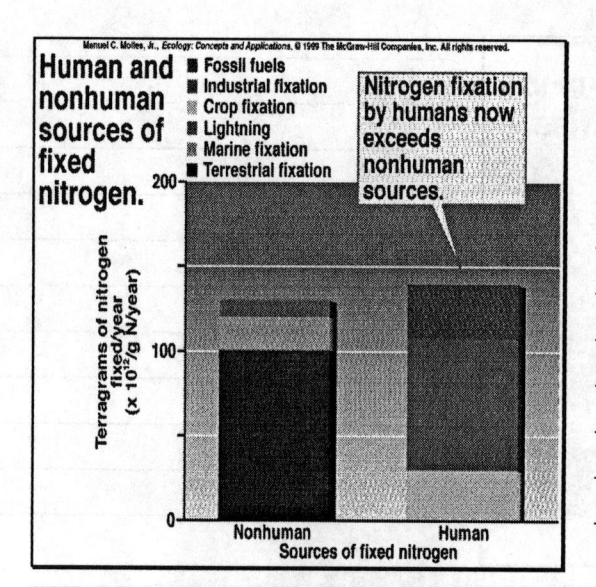

Human and nonhuman sources of fixed nitrogen.

- Fossil fuels
- Industrial fixation
- Crop fixation
- Lightning
- Marine fixation
- Terrestrial fixation

Nitrogen fixation by humans now exceeds nonhuman sources.

Sources of fixed nitrogen: Nonhuman, Human

Figure 20.14

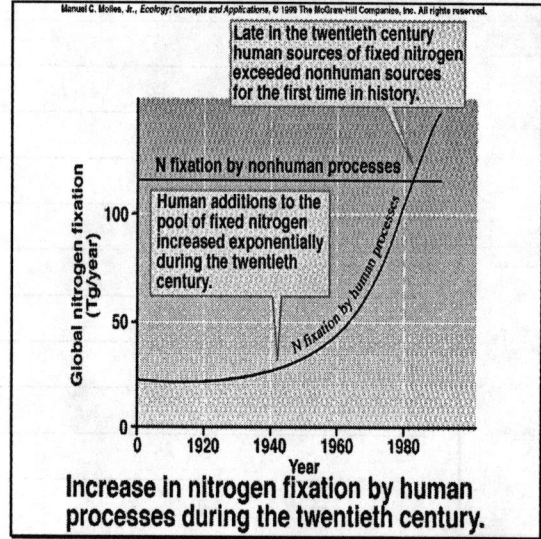

Late in the twentieth century human sources of fixed nitrogen exceeded nonhuman sources for the first time in history.

N fixation by nonhuman processes

Human additions to the pool of fixed nitrogen increased exponentially during the twentieth century.

N fixation by human processes

Increase in nitrogen fixation by human processes during the twentieth century.

Figure 20.15

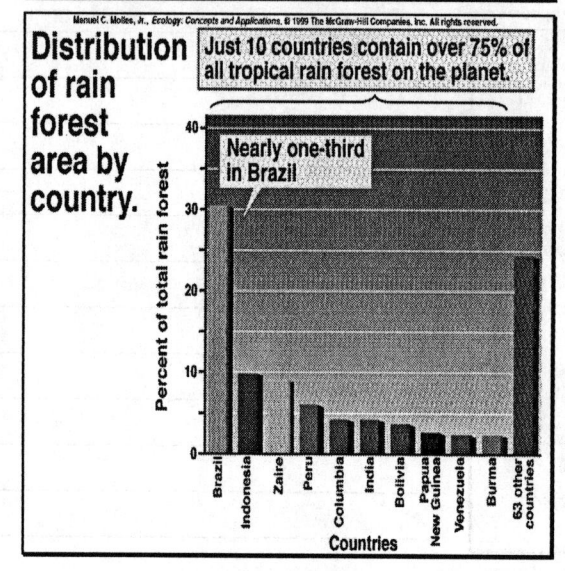

Distribution of rain forest area by country.

Just 10 countries contain over 75% of all tropical rain forest on the planet.

Nearly one-third in Brazil

Countries: Brazil, Indonesia, Zaire, Peru, Columbia, India, Bolivia, Papua New Guinea, Venezuela, Burma, 63 other countries

Figure 20.21

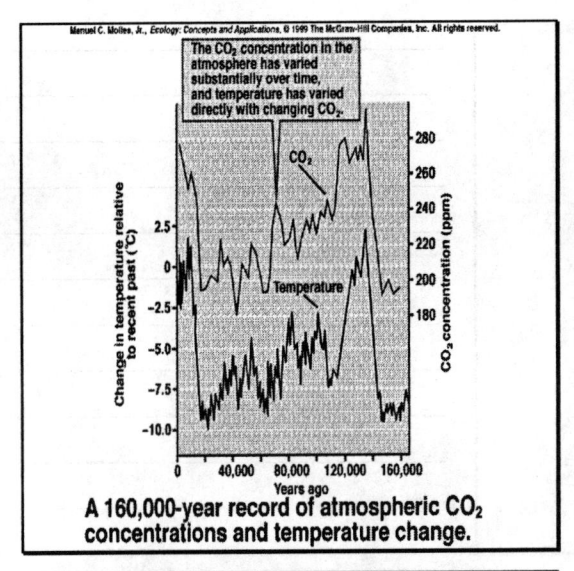

A 160,000-year record of atmospheric CO_2 concentrations and temperature change.

The CO_2 concentration in the atmosphere has varied substantially over time, and temperature has varied directly with changing CO_2.

Figure 20.22

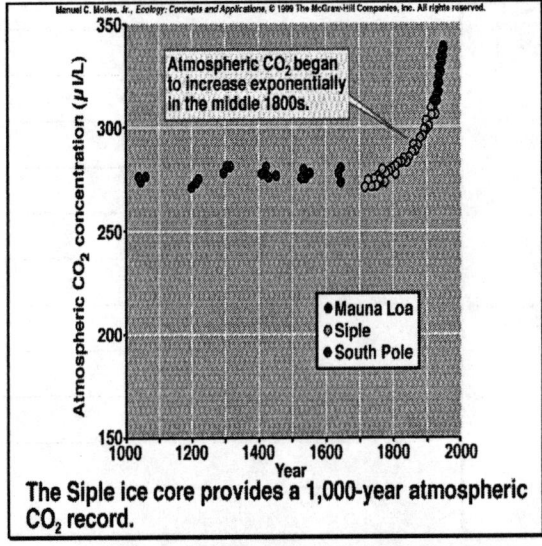

The Siple ice core provides a 1,000-year atmospheric CO_2 record.

Atmospheric CO_2 began to increase exponentially in the middle 1800s.

Figure 20.23

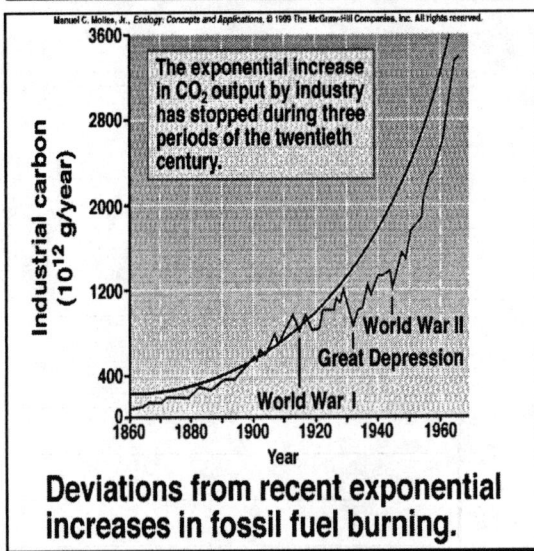

Deviations from recent exponential increases in fossil fuel burning.

The exponential increase in CO_2 output by industry has stopped during three periods of the twentieth century.

Figure 20.24

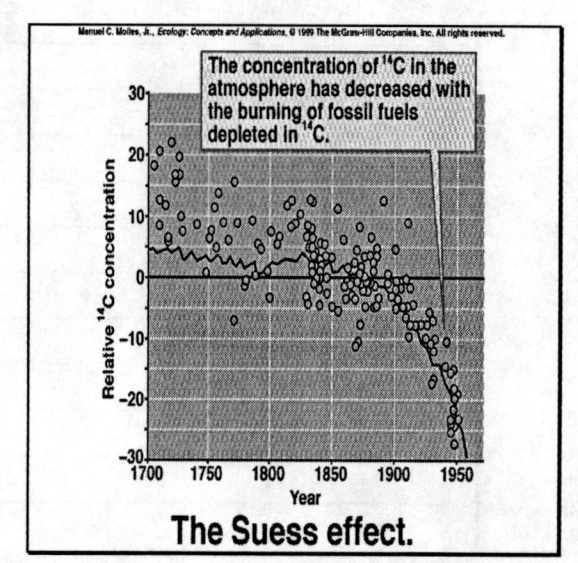

The concentration of ^{14}C in the atmosphere has decreased with the burning of fossil fuels depleted in ^{14}C.

The Suess effect.

Figure 20.25

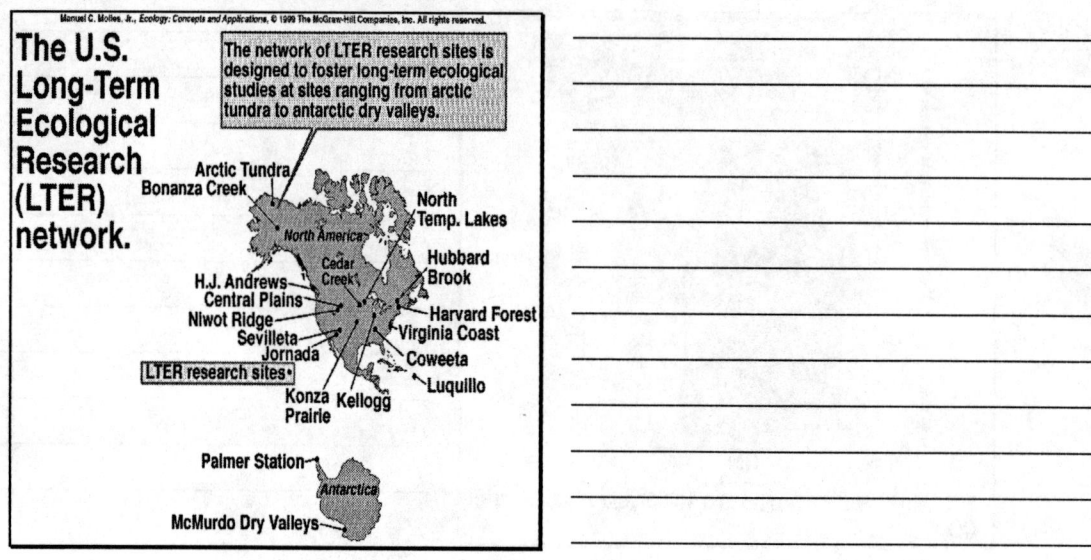

The U.S. Long-Term Ecological Research (LTER) network.

The network of LTER research sites is designed to foster long-term ecological studies at sites ranging from arctic tundra to antarctic dry valleys.